Mesenchymal Stem Cells in Cancer Therapy

Mesenchymal Stem Cells in Cancer Therapy

Khalid Shah
Massachusetts General Hospital,
Harvard Medical School,
Boston, Massachusetts, USA

AMSTERDAM • BOSTON • HEIDELBERG • LONDON
NEW YORK • OXFORD • PARIS • SAN DIEGO
SAN FRANCISCO • SINGAPORE • SYDNEY • TOKYO

ELSEVIER

Academic Press is an imprint of Elsevier

Academic Press is an imprint of Elsevier
The Boulevard, Langford Lane, Kidlington, Oxford, OX5 1GB, UK
225 Wyman Street, Waltham, MA 02451, USA

First published 2014

Notices
Knowledge and best practice in this field are constantly changing. As new research and
experience broaden our understanding, changes in research methods, professional practices, or
medical treatment may become necessary.

Practitioners and researchers must always rely on their own experience and knowledge in
evaluating and using any information, methods, compounds, or experiments described herein.
In using such information or methods they should be mindful of their own safety and the safety
of others, including parties for whom they have a professional responsibility.

To the fullest extent of the law, neither the Publisher nor the authors, contributors, or editors,
assume any liability for any injury and/or damage to persons or property as a matter of products
liability, negligence or otherwise, or from any use or operation of any methods, products,
instructions, or ideas contained in the material herein.

British Library Cataloguing in Publication Data
A catalogue record for this book is available from the British Library

Library of Congress Cataloging-in-Publication Data
A catalog record for this book is available from the Library of Congress

ISBN: 978-0-12-416606-6

For information on all Academic Press publications
visit our website at **store.elsevier.com**

This book has been manufactured using Print On Demand technology. Each copy is produced to
order and is limited to black ink. The online version of this book will show color figures where
appropriate.

CONTENTS

Genetically Engineered Mesenchymal Stem Cells: Targeted Foes of Cancer

1. INTRODUCTION

Currently, one in three women and one in two men in the United States will develop cancer in their lifetime [1]. Despite the considerable progress that has been made in nationally reducing cancer incidence and increasing patient survival, many common treatment options are in need of vast improvement and novel therapeutics are fervently desired. The most notable pitfalls of current treatments are the short half-life of a number of cancer-specific drugs, their limited delivery to some cancer types, and their adverse effects on vital noncancerous bodily tissues and functions. In response to such hindrances, many researchers have directed their studies to stem or progenitors cells. Most early work on stem cells was carried out with pluripotent embryonic stem cells (ESC) derived from inner cell mass (ICM) of blastocyst embryo. However, this, introduced a series of ethical problems in clinical applications. To avoid such ethical issues and create histocompatibility, adult stem cells have been isolated from different tissues including brain, heart, and kidney, and have emerged as attractive candidates to treat a wide range of diseases [2,3]. Also, new technologies have enabled tissue cells to become induced pluripotent stem cells (iPSC) [4]. Adult stem cells have been studied extensively and are already a successful source of FDA-approved treatments for a number of diseases, including Parkinson's disease and juvenile diabetes, due in part to their inherent regenerative properties [5]. Mesenchymal stem cells (MSC) are spindle-shaped, fibroblast-like multipotent stem cells that were first isolated and characterized by Friedenstein et al. in 1974 [6]. These cells are responsible for regeneration and cellular homeostasis in almost all tissues. The most intensely studied MSC are those derived from bone marrow and constitute nearly 10% of the hematopoietic stem cells (HSC) in number. In addition to bone marrow-derived MSC, these cells have also been derived from other sources, such as umbilical cord and adipose tissue. These MSC types have also been shown to expand rapidly in culture with sustained stable phenotype and differentiation potential into different mesenchymal lineages, such as fat, cartilage, and bone [7]. Recent studies have suggested that MSC are capable of differentiating into endodermal lineages as well [8,9]. In the last few years, several studies have demonstrated the *in vivo* characteristics of MSC. Sacchetti et al. and Crisan et al. revealed that MSC are likely linked to $CD146^+$ $CD45^-$ perivascular pericytes, which are capable of producing angiopoietin-1, an important molecule in HSC microenvironment [10,11]. Additionally, MSC have also been identified as $nestin^+$ cells in bone marrow, which play a critical role in constructing the HSC microenvironment [12]. This publication will shed light on the utilization of MSC in preclinical and clinical studies for cancer and also provide a perspective for their future clinical translation in cancer patients.

2. MSC SOURCES

Stem cells are the natural sources of embriogenetic tissue generation and continuous regeneration throughout adult life. In embryogenesis, cells from the ICM of the gastrula are known as "ESC" and their multilineage potential is generally referred to as pluripotent [13]. The gastrular ICM cells commence formation of the three germ layers: endoderm, mesoderm, and ectoderm, each committed to generating specified tissues of the forming body, and thus containing stem cells with more restricted potential than pluripotent stem cells [14]. Tissue-specific stem cells, such as mesenchymal stem cells (MSC) (mesoderm), HSC (mesoderm), and neural stem cells (ectoderm), have been identified as present and active for virtually every bodily tissue and are hierarchically situated between their germ layer progenitors and differentiated end-organ tissues [14]. Another stem cell type that has created new and compelling opportunities for cell therapies are induced pluripotent stem cells (iPSC). iPSC are created by inducing differentiated cells to express proteins that are specific to ESC. Resultantly, iPSC share many characteristics of ESC, including the ability to differentiate into cells of all organs and tissues. Somatic cells can be reprogrammed to a stem cell-like state by transferring their nuclear content into oocytes or by fusion with ESC, indicating that unfertilized eggs and ESC contain factors that can confer pluripotency to somatic cells [15,16]. iPSC are a novel and practical tool for human disease modeling and correction, and in theory could serve as a limitless stem cell source for patient-specific cellular therapies [16]. Although different stem cells offer a therapeutic potential, the preclinical studies discussed in this publication will mainly focus on MSC.

MSC can be isolated from adult human tissues, and have the capability for self-renewal and differentiation into mesenchymal lineages (osteocytic, chondrocytic, and adipogenic). The harvesting of MSC is generally free of ethical issues and is less invasive than other sources, such as neural stem cells. They can be expanded and manipulated *in vitro*, and subsequently regrafted. Following reimplantation, they have been found to suppress the immune system, reintegrate into tissue architecture, and give rise to progeny consisting of both stem cells and lineage restricted daughter cell types [2]. Most importantly, MSC exhibit potent pathotropic migratory properties, rendering them attractive for use in delivering targeted therapeutics to various cancers [2,3]. This treatment modality offers high site-specificity, and due to the immune suppressive nature of MSC they are protected from the host immune system, thus enabling them to assist in the long-term stability of delivered therapeutics; which also remedies potential problems related to limited biological drug half-life as drug secretion can be engineered to be continuous [2].

MSC have been successfully isolated from a number of organs (Figure 1) including brain, liver, kidney, lung, bone marrow, muscle, thymus, pancreas, skin, adipose tissue, fetal tissues, umbilical cord, Wharton's jelly, and placenta [17−21]. MSC possess the potential of converting to tissue types of other lineages, both within or across germ lines [22,23]. The highest degree of lineage plasticity has been imputed to bone marrow-derived MSC, which are capable of giving rise to virtually all cell types following implantation into early blastocysts and are relatively easy to handle *in vitro* [23,24]. Most of the preclinical studies to date have been performed with bone marrow-derived MSC, which might not be the most practical source available for the clinical settings. The harvesting of bone marrow requires an invasive procedure which yields a small number of cells, and the number, differentiation potential, and life span of bone marrow-derived MSC decline with patient age [25−27]. Two alternate sources for harvesting MSC that have received considerable attention in recent years are adipose tissue and umbilical cord blood (UCB). MSC derived from adipose have become a highly attractive alternative in recent years, largely due to the ease of tissue collection, high initial cell yields, and robust *in vitro*

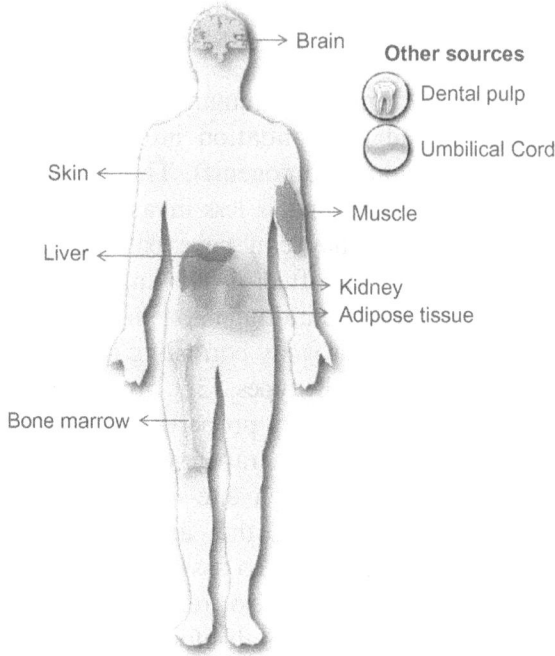

Figure 1 Sources of MSC.

proliferative capacity [28]. The expansion potential, differentiation capacity, and immunophenotype of MSC derived from adipose tissue are nearly identical to those isolated from bone marrow [26]. In addition, collection of stem cells from UCB or from the collagen-rich matrix surrounding the umbilical vasculature, Wharton's jelly, has been shown to also be a rich source of MSC [29]. As the cells are obtained after removal of the placenta, the collection procedure is entirely noninvasive and straightforward. Mononuclear cells can be separated and cultured from the cord blood, and cells in heterogenous adherent layer have been shown to have a fibroblastiod morphology, and express the same markers as bone marrow-derived MSC, namely CD13, CD29, CD49e, CD54, CD90 [30]. UCB-derived MSC expand at a higher rate as compared to bone marrow and adipose-derived (AD) MSC [26,31], which may be due in part to higher telomerase activity [32]. All three type of cells differentiate into osteocytes and chondrocytes [26,30,33,34] which is consistent with the properties of MSC.

Another source of MSC that has been explored and validated is adult dental pulp (DP). DP is a vascular connective tissue similar to mesenchymal tissue. The DP-derived stem cells have a phenotype similar to the adult bone marrow-derived MSC and these cells also express mesenchymal progenitor-related antigens SH2, SH3, SH4, CD166, and CD29 with a cellular homogeneity of 90–95%. Also, the DP and bone marrow-derived stem cell populations have a similar gene expression profile [35,36]. DP-MSC are derived from a very accessible tissue resource, which is further expandable by using deciduous teeth, and they possess stem cell-like qualities, including efficient self-renewal and multilineage differentiation. In addition, their capacity to induce osteogenesis could be of great clinical application in implantology [35,37]. DP-MSC could have potential clinical applications in autologous *in vivo* stem cell transplantation for calcified tissue reconstruction. Their proven immunomodulatory activity makes them suitable for suppression of T cell-mediated reaction in the setting of allogeneic bone marrow transplantation [35].

3. MSC HOMING AND MIGRATION

The first key step in MSC homing to tumors is their mobilization from the bone marrow and other organs. Endogenous MSC have been shown to mobilize from the bone marrow and other tissues to the peripheral blood under different injury conditions, normoxia, hypoxia, and inflammation [38,39]. The mechanisms by which MSC migrate across endothelium and home to the target tissues are not yet fully understood. However, it is known that a normal function of MSC is the ability to migrate to and repair wounded tissue. This wound healing property originates with migration toward inflammatory signals produced by the wounded environment [40]. Many of the same inflammatory mediators that are secreted by wounds are found in the tumor microenvironment and are thought to be involved in attracting MSC to these sites [41]. Extensive studies have shown that migration involves cytokines secreted by MSC that cooperate with G-protein coupled receptor (GPCR) and growth factor receptors, as well as the different cytokine/receptor pairs SDF-1/CXCR4, SCF-c-Kit, HGF/c-Met, VEGF/VEGFR, PDGF/PDGFR, MCP-1/CCR2, and HMGB1/RAGE [42]. Among these cytokine/receptor pairs, stromal cell-derived factor SDF-1 and its receptor CXC chemokine receptor-4 (CXCR4) are important mediators of stem cell recruitment to tumors. The importance of the interaction between secreted SDF-1 and cell surface CXCR4 for stem cell migration has been displayed by experiments in which the activity of either the receptor or the cytokine has been inhibited [43–45]. The blockade of both CXCR4 and SDF-1 *in vivo* in diseased mice has been found to markedly reduce the migration of transplanted stem cells toward tumor foci and regions of demyelination, indicating that SDF-1/CXCR4 signaling is essential for effective pathotropism of therapeutic stem cells [43]. Recent data with inhibitors to chemokine receptor CXCR4 and TGFβ receptor suggest that endogenous MSC homing to tumors, differentiation to myofibroblasts, and/or survival require CXCR4 [46]. Apart from CXCR4, MSC are known to express a broad range of chemokine receptors including CXCR1, CXCR3, CXCR4, CXCR5, CXCR6, chemokine (C–C motif) receptor (CCR)-1, CCR2, CCR3, CCR4, CCR5, CCR9, and others. Recent studies have shown that chemokines such as CXCL12, CXCL13, CXCL16, and their receptors can enhance the bidirectional migration of MSC to and from the bone marrow niche [47]. It is known that specific chemokines/receptor pairs are involved in the unidirectional migration of MSC. CXCL16 (ligand for CXCR6) is effective in the homing of MSC into the bone marrow, while CCL22 (ligand for CCR4) has the strongest chemotactic effect in mobilizing MSC from the bone marrow into the circulation [47]. Both CXCL16 and CCL2 are known to be expressed in tumor tissues such as lung carcinoma, hepatocellular carcinoma (HCC) colorectal cancer [48], and malignant brain and ovarian cancers [49,50], and it is plausible that they have a role to play in the migration of MSC into tumors.

Cell receptor transactivation is a thoroughly researched mechanism thought to be involved in migration, which involves the activation of one receptor by another. It not only plays an important physiological role in processes such as cellular migration and apoptosis, but its deregulation can cause pathological states such as cancer. Transactivation of various growth factor receptors, including epidermal growth factor receptor (EGFR), by GPCRs has been documented in multiple cellular model systems, hence demonstrating the potential role of GPCR in tumor tropism via receptor transactivation [51–53]. A commonly reported mechanism of receptor transactivation involves the activation of membrane-tethered growth factors, such as EGFR, through direct interaction with GPCR, such as CXCR4 [54]. This process is assisted by matrix metalloproteinases (MMPs) such as MMP-2 and MMP-9 [55], which are proteinases required to proteolytically process precursor proteins such as adhesion molecules, growth factors, cytokines, and their receptors. In a study by De Becker and colleagues, it was reported that the migration of MSC through bone marrow endothelium was mediated by MMP-1 and tissue inhibitor of metalloproteinase-3 [56]. In another study, elevated levels of MMP-2 were observed to be responsible for C1q complement protein-mediated migration of cord blood-derived MSC toward injured tissue and organ [57]. A recent report on MSC behavior indicates that MSC are attracted to sites of irradiation, and that local irradiation might promote specificity of MSC migration and engraftment [58]. Although these findings are not surprising in the light of general stem cell tropism for injured tissues, they do stress the potential synergism between radiotherapy and tumor-specific MSC targeting in the clinical arena.

Besides targeting the tumor main burden, MSC and other stem cell types have been shown to track tumor metastases and small intracranial microsatellite deposits of different tumor types, and effectively treat these by either the factors released by stem cells or in loco expression of tumoricidal transgenes that they have been engineered with [59–61]. These findings provide a strong rationale for the development of therapies that capitalize on the tumoritropic properties of MSC by engineering them into carriers for antitumor therapy.

4. MSC FOR TUMOR THERAPY

Unmodified MSC have been shown to have antitumor effects both *in vitro* and in different mouse models of cancer. This is attributed to the factors released by MSC that have antitumor properties that can reduce the proliferation of glioma, melanoma, lung cancer, hepatoma, and breast cancer cells [62–65]. Human bone marrow-derived MSC injected intravenously in a mouse model of Kaposi's sarcoma were shown to home to sites of tumorigenesis and potently inhibit tumor growth [66]. MSC have also been shown to have antiangiogenic effect both *in vitro* and in mouse models of melanoma [67]. Direct injection of MSC into subcutaneous melanoma bearing mice induced apoptosis and abrogated tumor growth [67]. Human umbilical cord blood (hUCB)-derived MSC have been used as naïve cells for the treatment of glioblastoma multiforme (GBM). UCBSC enriched in CD44 and CD133 cells cocultured with GBM cells underwent apoptosis [68]. Treatment of glioma cells with hUCBs also inhibited focal adhesion kinase (FAK)-mediated angiogenesis [69], upregulated PTEN in glioma cells, and in the nude mice tumors and downregulated Akt and PI3K signaling pathway molecules thereby resulting in the inhibition of migration as well as wound healing property of the glioma cells [70]. This simultaneously resulted in downregulation of XIAP activating caspase-3 and caspase-9 to trigger apoptosis in glioma cells [71]. In another study, UCB-MSC were shown to inhibit glioma growth, reduce neovascularization, and decrease cyclin D1 protein expression *in vivo* [72]. In a recent study, comparative study UCB-MSC and not AT-MSC were shown to inhibit GBM growth via tumor necrosis factor (TNF)-related apoptosis-inducing ligand (TRAIL) [73]. MSC have been genetically modified mainly to introduce and over-express target exogenous genes for expression/secretion of a desired therapeutic factor for targeted treatment of different cancer types (Figure 2).

4.1 MSC Delivery of Interleukins

Interleukins (IL) are cytokines that regulate inflammatory and immune responses and are known to have antitumor properties via direct tumoricidal effects or positive modulation of the endogenous immune system [74]. However, a lack of tumor-targeted delivery has hindered

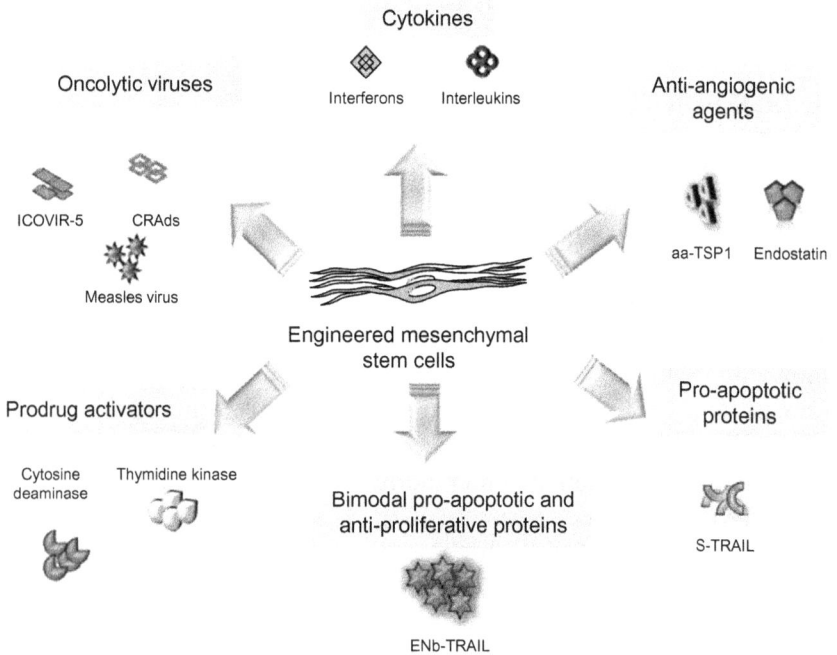

Figure 2 Transgene strategies for MSC-based therapeutics.

their application for cancer therapy [75]. Resultantly, the delivery of IL by MSC has been explored. MSC engineered to express IL have been utilized to improve the anticancer immune surveillance by activating cytotoxic lymphocytes and natural killer cells [74]. MSC expressing IL-12 have been shown to prevent metastasis into the lymph nodes and other internal organs as well as increased tumor cell apoptosis in mice bearing pre-established metastases of melanoma, breast, and hepatoma tumors [76]. In two previous studies, MSC expressing IL-12 were shown to have antitumor effects in mice bearing renal cell carcinomas [77] and cervical tumors [78]. Both studies revealed sustained expression of IL-12 and interferon (IFN)-γ in sera and tumor sites. Furthermore, in a 2011 study, hUCB-derived MSC were successfully employed as vehicles to deliver IL-12 to aid in the treatment of malignant glioma [79]. Similarly, the transplantation of IL-18 secreting MSC was previously shown to enhance T cell infiltration and long-term antitumor immunity in mice bearing noninvasive and invasive gliomas [80]. In another study, hUCB-MSC were engineered to express IL-21 and were shown to have therapeutic efficacy in mice bearing ovarian cancer xenografts [81]. In a recent study, Zhang et al. evaluated the effects of IL-24 delivered by MSC as a therapeutic approach for lung cancer. Human umbilical cord-derived MSC (HUMSC) engineered to deliver secretable IL-24, were shown to inhibit the growth of lung cancer cells by induction of apoptosis and cell cycle arrest [75]. This study also demonstrated that injection of MSC secreting IL-24 significantly suppressed xenograft tumor growth and had antiangiogenic effects both *in vitro* and *in vivo* [75]. The results of these experiments indicate that MSC-delivered IL have the potential to be used as an alternative strategy for cancer therapy.

4.2 Interferons

IFN-β has been shown to have anti-proliferative and pro-apoptotic effects [82–84]. However, its *in vivo* therapeutic efficacy has been limited due to toxicity associated with systemic administration. A potential solution to this problem is to engineer human MSC to express IFN-β, a strategy that has been used for targeted delivery to metastatic breast and melanoma models [85,86], gliomas [44], and lung metastasis [44,87,88]. Recently, Dembinski and colleagues intraperitoneally injected mice bearing ovarian xenografts with IFN-β expressing MSC which resulted in the complete eradication of tumors in 70% of treated mice and an increased survival in others [89]. This study confirmed

MSC's potential as a targeted delivery vehicle for the intratumoral production of IFN-β. In 2012, Wang et al. [90] investigated the use of MSC engineered to produce IFN-β. Intravenously injected MSC-IFN-β significantly reduced prostate tumor weight and increased animal survival compared with controls [90]. In another recently published study, amniotic fluid (AF)-derived MSC were investigated to transport IFN-β to the region of neoplasia in a bladder tumor model. A significant inhibition of tumor growth as well as prolonged survival of mice was observed in the presence of AF-MSC-IFN-β [91].

Earlier studies include the work of Studeny and colleagues, who engineered human adult MSC expressing INF-β and showed their *in vivo* efficacy against solid melanomas in nude mice [85]. In 2005, Nakamizo and colleagues investigated the antitumor effects of MSC expressing INF-β in central nervous system (CNS) tumors, by evaluating whether human MSC could still track murine brain tumors when administered through the bloodstream [44]. By manipulating MSC to secrete INF-β, this tropism could be exploited for antitumor effect, as *in vivo* administration of hMSC-INF-β resulted in significantly enhanced murine survival. In a related study, Ren et al. [88] evaluated the potential of MSC expressing IFN-β in a model of prostate cancer lung metastasis. Targeted homing of MSC producing IFN-β was seen at sites of tumor in the lungs with established pulmonary metastases, and this resulted in suppression of tumor growth.

The antitumor effects of a multifunctional regulatory cytokine, IFN-α, has also been investigated. IFN-α is frequently used as a supplementary therapeutic agent to eradicate micrometastatic deposits in patients with a high risk of systemic recurrence [92,93]. Sartoris and colleagues found that MSC expressing IFN-α can be efficiently delivered inside the tumor microenvironment of a mouse plasmacytoma model [94]. Subcutaneous administration of MSC-IFN-α significantly impeded the tumor growth *in vivo* and prolonged the overall survival of the mice by inducing apoptosis in tumor cells and by reduction in tumor vessel density [94]. A similar study in 2006 explored the therapeutic efficacy of MSC expressing IFN-α for the treatment of lung metastasis in a mouse model of metastatic melanoma. The systemic administration of MSC expressing IFN-α reduced the growth of melanoma cells and significantly prolonged survival due to an increase in tumor cell apoptosis and a decrease in blood vasculature [87].

4.3 Prodrugs

The conversion of nontoxic prodrugs into toxic antimetabolites, known as prodrug activation schemes, is available for selective killing of tumor cells. Herpes simplex virus (HSV)-1 thymidine kinase (TK), cytosine deaminase (CD), and carboxyesterase genes, confer sensitivity to ganciclovir (GCV), 5-fluorocytosine (5-FC), and camptothecin-11 (CPT-11), respectively. These systems are currently being evaluated in clinical trials for selective elimination of tumor cells [95]. Human MSC derived from bone marrow and adipose tissues, as well as neural stem cells, have been found to be effective vehicles for gene directed enzyme prodrug therapy [96].

Activated prodrugs that are not toxic to MSC and have a bystander tumor-killing effect were initially explored using CD, which can convert the nontoxic "prodrug" 5-FC to the drug, 5-fluorouracil (5-FU), a chemotherapeutic agent that can readily diffuse out of the producer stem cell and into surrounding cells and is selectively toxic to rapidly dividing cells [59]. Miletic and colleagues found that MSC that was engineered to express HSV-TK and then injected into the tumor or the vicinity of the tumor, infiltrated solid parts as well as the border of glioma models in rats, and ultimately showed high therapeutic efficacy by significant reduction of tumor volumes through bystander-mediated glioma cell killing [97]. In 2012, Choi et al. studied human adipose tissue (hAT)-derived MSC and prodrug gene therapy against brain stem gliomas in rat models. hAT-MSC were modified to express rabbit carboxylesterase (rCE) enzyme, which can efficiently convert the prodrug CPT-11 into the active drug SN-38 (7-ethyl-10-hydroxycamptothecin). A significant increase in the survival time of rats treated with hAT-MSC.rCE and CPT-11 was observed than in rats treated with CPT-11 alone, demonstrating the therapeutic potential of MSC as cellular vehicles for prodrug gene therapy in gliomas [98]. Song and colleagues studied bone marrow-derived MSC infected with HSV-TK in mouse models of prostate cancer. MSC-HSV-TK significantly inhibited the growth of prostate cancer xenografts in the presence of GCV. Additionally, the MSC-HSV-TK exerted a significant antitumor effect in an animal model of metastastic RIF-1 (fibrosarcoma) tumor in the presence of prodrug GCV and proved to not cause any harmful side effects *in vivo* [99]. In a past study, the ability of AT-MSC engineered to express the suicide gene cytosine deaminase::uracil phosphoribosyltransferase (CD::UPRT) was explored in mouse models of prostate cancer [17]. CD::UPRT converts the relatively nontoxic 5-FC into the highly toxic antitumor 5-FU. Therapeutic AT-MSC expressing CD::UPRT proved effective in significantly inhibiting prostate cancer tumor growth after intravenous administration in mice bearing tumors and treated with 5-FC [100].

In a recent study, we have developed an efficient stem cell-based therapeutic strategy that simultaneously allows killing of GBM tumor cells and assessment and eradication of stem cells (SC) post-tumor treatment. MSC engineered to co-express the prodrug converting enzyme, HSV-TK and secretable TRAIL (S-TRAIL), induced caspase-mediated GBM cell death, and showed selective MSC sensitization to the prodrug GCV (Figure 3). A significant decrease in tumor growth and a subsequent increase in survival were observed when mice bearing a highly aggressive GBM were treated with MSC co-expressing S-TRAIL and HSV-TK. Furthermore, the systemic administration of GCV post-tumor treatment selectively eliminated therapeutic MSC expressing HSV-TK *in vitro* and *in vivo*. These findings demonstrate the development and validation of a novel therapeutic strategy that has implications in translating stem cell-based therapies in cancer patients [101].

4.4 Oncolytic Viruses

The ability of oncolytic viruses (OV) to selectively replicate in and destroy tumor cells, while sparing healthy cells, makes them strong candidates for anti-tumor therapy [102,103]. Due to clearance of the virus by host defense mechanism and spurious targeting of noncancer tissues through the bloodstream, the systemic administration of OV is often inefficient [104]. Cell-mediated OV delivery could shield the virus from host defenses and direct them toward tumors. Different stem cell types, including MSC, have been used as host cells for the replication, transportation, and local release of intact, conditionally replicating oncolytic adenoviruses (CRAd) [105]. Human MSC were shown to support replication of adenovirus bearing TK and to have bystander effect against different cancer cell lines [106]. When administered intravenously into murine models of solid ovarian cancer, CRAd-charged MSC resulted in significantly enhanced antitumor effect and extended survival as compared to direct delivery of CRAd [107].

MSC have also been employed to deliver adenovirus which subsequently infected and replicated within malignant cells and eradicated the tumors [107,108]. MSC have been utilized to deliver CRAd in a mouse

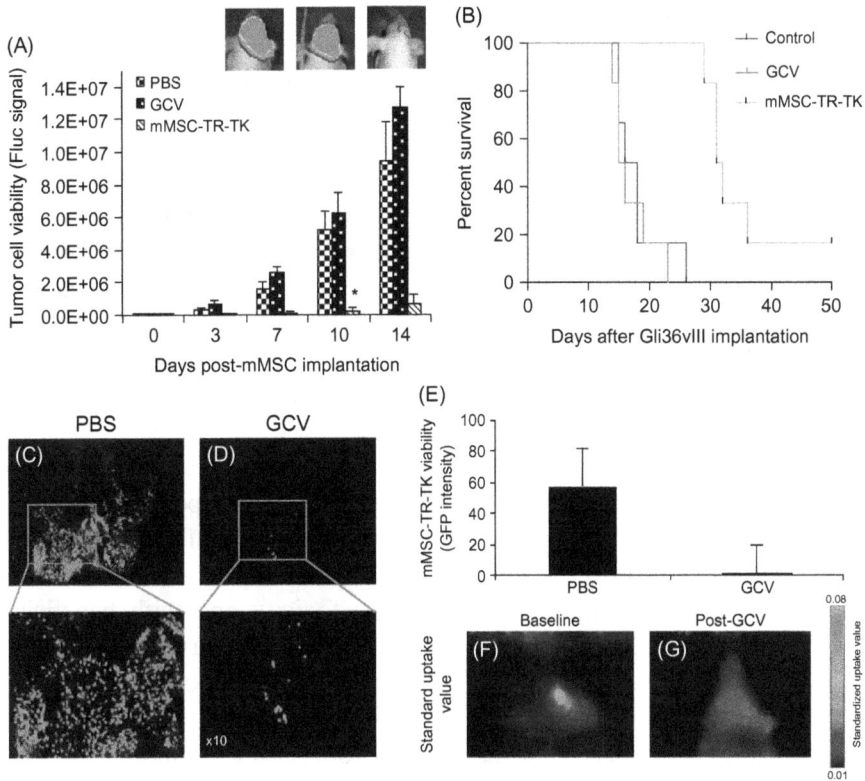

Figure 3 MSC co-expressing S-TRAIL and HSV-TK have antitumor effects and their fate can be imaged in vivo. Mice bearing highly malignant Gli36vIII-FmC GBMs were treated intratumorally with MSC-TR-TK, control PBS, or GCV intraperitoneally and imaged for Fluc activity. (A) Plot shows the relative mean bioluminescent signal intensity after quantification of in vivo images. One image from a representative mouse of each group at day 14 post-mMSC implantation is shown. Bars, + SE. (B) Kaplan–Meier survival curves of Gli36vIII-FmC bearing mice treated with PBS, GCV, or MSC-TR-TK. (C–E) Gli36vIII-FmC bearing mice were intratumorally implanted with MSC-TR-TK and 14 days later treated with GCV or PBS. Photomicrographs (original magnification, ×4 upper panel and ×10 lower panel) from brain sections of PBS (C) or GCV (D) groups post 18 days after GCV treatment are shown. Quantification of MSC-TR-TK viability measured by GFP intensity from the brain sections (E). In the same experiment above, MSC-TR-TK were imaged by PET using 18F-FHBG pre- (F) and post- (G) 10 days with GCV treatment. Representative images of a mouse from each group are shown. Adapted from Ref. [101] with permission.

model of intracranial malignant glioma [109]. CRAd-loaded MSC resulted in efficient adenoviral infection of distant glioma cells confirming the ability of MSC as carriers for oncolytic adenoviral vectors for the treatment of malignant glioma. In a previous study, delivery and efficacy of oAV, Delta24-RGD by human MSC has been assessed in mouse models of glioblastomas [110]. MSC-Delta24 that were injected into the carotid artery of mice harboring orthotopic glioma xenografts, selectively localized to glioma xenografts and released

Delta24-RGD, which subsequently infected glioma cells, inhibited glioma growth and resulted in eradication of tumors with a significant increase in the median survival of treated animals as compared to controls.

The therapeutic approach of engineered adenoviruses has been studied for the treatment of metastatic tumors; however, systemic delivery of these oncolytic adenoviruses lacks metastatic targeting ability [111]. The tumor stroma engrafting property of MSC may allow them to be used as cellular vehicles for targeted delivery. Garcia-Castro and colleagues explored the safety and efficacy of infusing autologous MSC infected with ICOVIR-5, a new oncolytic adenovirus, for the treatment of metastatic neuroblastoma. In the study, four children with metastatic neuroblastoma that was resistant to frontline therapies, received several doses of autologous MSC carrying ICOVIR-5 [111]. The children's tolerance to the treatment was excellent, and a complete clinical response was documented in one case, with the child in remission 3 years after the therapy. These results indicate that MSC can deliver oncolytic adenoviruses to metastatic tumors with very low systemic toxicity and with beneficial antitumor effects. Recently, MSC have been shown to serve as carriers to deliver oncolytic measles virus (MV) to ovarian tumors [112]. MSC obtained from ovarian cancer patients migrated toward primary ovarian cancer samples in chemotaxis assays and to ovarian tumors in athymic mice. Using noninvasive single photon emission tomography—computed tomography (SPECT—CT) imaging, a rapid co-localization of MV-infected MSC to the ovarian tumors was observed within 5—8 minutes of their intraperitoneal administration. Furthermore, MV-infected MSC, but not virus alone, significantly prolonged the survival of measles immune ovarian cancer bearing animals. In a recent study, the potent antitumor activity of systemically delivered MV-infected bone marrow-derived MSC was explored in human HCC tumors in SCID mice passively immunized with human neutralizing antibodies against MV. MSC infected with MV homed to the HCC tumors and resulted in a significant inhibition of tumor growth in both measles antibody-naïve and passively immunized SCID mice [113]. These studies confirm the feasibility of using MSC as carriers for oncolytic MV therapy.

4.5 Antiangiogenic Agents

Tumor angiogenesis represents a way for cancer cells to function and thrive for self-sustained growth [114]. Inhibition of tumor-induced angiogenesis may restrict tumor growth and metastasis. However, long-term systemic delivery of angiogenic inhibitors is associated with toxicity, as well as pruning away the vessels, which allows tumors to become more chemo-resistant due to inadequate delivery of drugs [115]. The utility of MSC as vehicles for antiangiogenic therapeutics has been studied, as they exhibit a tropism to cancer tissue and may deliver anti-angiogenic agents without adverse side effects [116]. It has also been determined that tumor-mediated angiogenesis results from a dysregulation of both pro-angiogenic and antiangiogenic factors as well as by various growth factors and molecules of the extracellular matrix, which has led to the study and utilization of targeted antiangiogenic therapy to inhibit tumor proliferation and neovascularization [117,118].

Recently, Zheng et al. studied endostatin, which is an important endogenous inhibitor of tumor vascularization that has been widely used in antiangiogenic therapy for various cancers [119]. Human placenta-derived mesenchymal stem cells (hpMSC) were engineered to deliver endostatin via adenoviral transduction and were injected into nude mice. The hpMSC expressing the human endostatin gene demonstrated preferential homing to the tumor site and significantly decreased the tumor volume without apparent systemic toxic effects. These observations were associated with significantly decreased blood vessel formation, tumor cell proliferation, and increased tumor cell apoptosis [119].

In a previous phase II clinical study, it was found that the delivery of antiangiogenic drugs through vasculature normalized the abnormal structure and function of the blood vessels and resulted in reduction of tumor-associated vasogenic brain edema in most patients [120]. The vessel normalization is associated with a significant decrease in their mean vessel diameter and permeability [121,122] and increased pericyte coating of small vasculature [123]. MSC are known to localize to tumor vasculature upon intratumoral implantation, thus offering increased abilities for targeting particularly vascularized tumors [124].

4.6 Pro-apoptotic Proteins

The delivery of pro-apoptotic protein, TRAIL, via SC offers a promising on-site delivery approach toward tumor cell killing [17]. TRAIL is an endogenous member of the TNF ligand family that binds to its death domain containing receptors Dr4 and Dr5 and induces apoptosis via activation of caspases, preferentially in cancer cells while sparing most other cell types [125]. A number of studies have shown the therapeutic efficacy of different adult stem cell types, including MSC, engineered to express TRAIL in either cell lines or mouse models of colorectal carcinoma [126], gliomas [127–129], lung, breast, squamous, and cervical cancer [130], resulting in induction of apoptosis and a subsequent reduction of tumor cell viability. Grisendi and colleagues found that AD MSC armed with TRAIL targeted a variety of tumor cell lines *in vitro*, including human cervical carcinoma, pancreatic cancer, and colon cancer. When injected into mice, AD MSC-TRAIL migrated to tumors and induced apoptosis in tumor cells, without significant toxicities to normal tissues [131]. In a similar study, AD MSC engineered to express TRAIL was found to exhibit anti-myeloma activities and significantly induce myeloma cell death *in vitro* [132].

Since TRAIL is a type II membrane protein and its release into the microenvironment requires additional cleavage from its cell membrane anchoring site, our lab previously worked on redesigning the TRAIL protein in order to engineer truly paracrine TRAIL-secreting cells. We have designed a secretable version of TRAIL that consists of a fusion between the extracellular domain of TRAIL and the extracellular domain of the hFlt3 ligand, which binds to the Flt3 tyrosine kinase receptor [17]. The re-engineered recombinant protein named S-TRAIL is efficiently secreted into the producer cell's immediate microenvironment and exhibits higher cytotoxicity on tumor cells than the native TRAIL protein [60,133,134]. In a previous study, we demonstrated that human MSC are resistant to TRAIL-mediated apoptosis and when engineered to express S-TRAIL, induce caspase-mediated apoptosis in established glioma cell lines as well as glioblastoma stem cells (GBSC) *in vitro*. Using highly malignant and invasive human glioma models generated from human GBSC and employing real-time imaging, we have shown that MSC-S-TRAIL migrate extensively to tumors in the brain and have profound antitumor effects *in vivo*. This study demonstrates the efficacy of therapeutic S-TRAIL and the potential of human MSC to be used as delivery vehicles targeting GBSC *in vivo* [60].

5. SYNERGISTIC APPROACHES UTILIZING MSC-BASED THERAPEUTICS WITH OTHER ANTITUMOR AGENTS

Given the heterogeneity of tumors in general, it is unlikely that any one effective strategy will provide a satisfactory treatment regimen for tumors. The advent of molecular theragnostics and personalized medicine might largely remedy the differences in nature and therapeutic resistance between different tumors [2,135], but cannot provide adequate answers to the existence of profound intratumoral heterogeneity as is observed, for instance, in gliomas [136]. A practical approach would be to combine distinct therapeutic targets, such as those involved in tumor cell growth, apoptosis, and the proliferation of tumor-associated vasculature to fully eradicate different tumor types. Given that 50% of tumor lines are resistant to TRAIL, overcoming TRAIL resistance in aggressive tumors, such as GBM, and understanding the molecular dynamics of TRAIL-based combination therapies are critical to using TRAIL as a therapeutic agent. In a recent study, we engineered human MSC to express S-TRAIL and assessed the ability of MSC-S-TRAIL-mediated medulloblastoma (MB) killing when combined with a small molecule inhibitor of histone deacetylase, Ms-275, in TRAIL-sensitive and -resistant MB *in vitro* and *in vivo* [137]. In TRAIL-resistant MB, we showed upregulation of Dr4/5 levels when pretreated with Ms-275 and a subsequent sensitization to MSC-S-TRAIL-mediated apoptosis. Using intracranially implanted MB and MSC lines engineered with different combinations of fluorescent and bioluminescent proteins, we found that MSC-S-TRAIL has significant antitumor effects in mice bearing TRAIL-sensitive and Ms-275 pretreated TRAIL-resistant MBs [137]. In another study, a therapeutic combination of the lipoxygenase inhibitor MK886 and TRAIL-secreting human MSC was explored. MK886 effectively increased the sensitivity to TRAIL-induced apoptosis via upregulation of the death receptor 5 and downregulation of the antiapoptotic protein survivin in human glioma cell lines and in primary glioma cells. *In vivo* survival experiments and imaging analysis in orthotopic xenografted mice showed that MSC-based TRAIL gene delivery combined with MK886 into the tumors had greater therapeutic efficacy than single-agent treatment [138]. We have also designed additional supplementary treatments, like microRNA-21 (miR-21) inhibitors [139] and novel PI3-kinase/mTOR inhibitor, PI-103 [140] to augment the antitumor effect of different stem cell-mediated S-TRAIL therapy *in vivo*. A similar study has demonstrated that the combined approach using systemic MSC-mediated delivery of TRAIL together with XIAP inhibition suppresses metastatic growth of pancreatic carcinomas [141]. These findings offer a preclinical rationale for application of mechanism-based systemically delivered anti-proliferative agents and novel stem cell-based proapoptotic therapies to improve treatment of malignant tumors.

In addition to molecular approaches, current clinical treatment regimens such as local radiotherapy might be suited for enhancing stem cell therapy, as their effects on irradiated tissue seem to additionally promote the homing of transplanted stem cells [58]. Past studies have revealed that tumor irradiation enhances the tumor tropism of hUCB-derived MSC by increased IL-8 expression on glioma cells [142]. The sequential treatment with irradiation followed by TRAIL-secreting UCB-MSC synergistically enhanced apoptosis in glioma cells by up-regulating expression of Dr5 and subsequently inducing caspase activation. *In vivo* survival experiments in orthotopic xenografted mice showed that MSC-based TRAIL gene delivery to irradiated tumors had greater therapeutic efficacy than a single treatment. These results suggest that clinically relevant tumor irradiation increases the therapeutic efficacy of MSC-TRAIL by increasing tropism of MSC and TRAIL-induced apoptosis, which might be a more useful therapeutic strategy for treating tumors in general and gliomas in particular.

In a recent study, murine MSC expressing TRAIL were tested in combination with conventional chemotherapeutic drug treatment in colon cancer models [143]. A significant decrease in tumor volumes was seen when mice bearing HCT116 colorectal cancer xenografts were co-treated with 5-FU and systemically delivered MSC-S-TRAIL. This antitumor effect was protein 53 (p53) independent and was mediated by TRAIL-receptor 2 (TRAIL-R2) upregulation, demonstrating the applicability of this approach in p53-defective tumors. In another study, Yulyana and colleagues combined the tumor selectivity of TRAIL and tumor-homing properties of MSC with gap junction (GJ) inhibitory effect of carbenoxolone (CBX) to target orthotopic gliomas [144]. *In vitro* studies revealed that CBX enhanced TRAIL-induced apoptosis through upregulation of death receptor 5, blockade of GJ intercellular communication, and via downregulation of connexin 43. Dual arm therapy using MSC-TRAIL and CBX prolonged the survival of treated mice by $\sim 27\%$ when compared with the controls in an intracranial glioma model. The enhanced efficacy of MSC-TRAIL in combination with either 5-FU or CBX coupled with the minimal cytotoxic nature of these agents suggests their potential clinical implementation.

The ideal stem cell-based combination therapies would entail SC expressing multi-targeted molecules, such as when TRAIL and/or other biomolecules are produced by stem cells. In a recent study, we demonstrated the simultaneous expression of a multifunctional biomolecule by stem cells, specifically, a fusion of EGFR antagonist (EGFR-Nb) and TRAIL. We found that the expression of EGFR-Nb and TRAIL from a single stem cell source caused enhanced killing of tumor cells as opposed to expression of each molecule from stem cells separately [145]. MSC were engineered to co-express dodecameric TRAIL and HSV-TK (MSC/dTRAIL-TK) and their antitumor effects were assessed in an experimental lung metastasis model [146]. MSC/dTRAIL-TK treatment followed by GCV administrations significantly decreased the number of tumor nodules in the lung as compared to MSC/dTRAIL or MSC/TK treatment alone, and resulted in 100% survival of tumor bearing mice after three injections. Recently, another promising double-target stem cell-based therapeutic system was explored for non-Hodgkin's lymphoma (NHL). HUMSC engineered to express scFvCD20-S-TRAIL which contains a CD20-specific single chain Fv antibody fragment (scFv) and a soluble TRAIL, were shown to significantly inhibit tumor growth in beating lymphoma xenografts [147].

6. ENCAPSULATED MSC FOR THERAPY

Cell encapsulation technology refers to immobilization of cells within biocompatible, semipermeable membranes. The encapsulation of cells, instead of therapeutic products, allows the delivery of molecules of interest for a longer period of time as cells release these molecules continuously. In addition, cells can be engineered to express any desired protein *in vivo* without the modification of the host's genome [148]. Cell encapsulation presents an important advantage as compared to encapsulation of proteins, as the former allows a sustained and controlled delivery of therapeutic molecules at a constant rate giving rise to greater physiological concentrations [148]. Due to their ability to provide a physiologic environment that promotes cell survival and prevent immune response while permitting easy *in vivo* transplantation and cell retention, biodegradable hydrogels and synthetic extracellular matrix (sECM) to encapsulate stem cells have been utilized in a variety of rodent models [149,150]. A number of different biomaterials such as hyaluronic acid, alginate, agarose, and other polymers have been used for encapsulation. In past studies, sECM acted as the necessary biomechanical substrate for endogenous neuroregeneration in models of intracerebral hypoxia-ischemia and traumatic spinal cord injury, by increasing their stem cell viability and promoting differentiation into neurons [151–153]. Recently, it was discovered that MSC encapsulated in fibrinogen–alginate microcapsules possessed a significantly increased survival as compared to un-encapsulated cells [154]. In another recent study, Reagan et al. demonstrated the utility of a scaffold-based delivery system for sustained therapeutic MSC release and their ability to express genetically introduced therapeutic TRAIL. MSC expressing full length TRAIL under a doxycycline inducible promoter were encapsulated in porous, biocompatible silk scaffolds and administered to mice using different administration routes. Encapsulated MSC-TRAIL successfully decreased bone and lung metastasis, whereas liver metastasis decreased only with tail vein administration routes upon doxycycline administration [155].

Stem cell encapsulation is an important prospect for the treatment of GBM. The recurrence rates of GBM and associated patient mortality are nearly 100%, which is largely attributed to inefficient delivery of many therapeutic molecules to brain tumor cells, due to the blood–brain barrier (BBB) [156] and vascular dysfunction in the tumor [114]. One of the approaches to overcoming drug delivery problems to intracranial tumors is to develop on-site means to deliver novel tumor-specific agents. However, in order to effectively deliver such therapeutic agents, methods must be developed to introduce stem cells into the resection cavity while preventing the rapid "washout" of a significant number of cells by cerebrospinal fluid (CSF). Additionally, it is critical to allow efficient secretion of anti-GBM therapies and retain the ability of stem cells to migrate from the resection cavity into the parenchyma toward invasive tumor deposits. In the study from Kauer et al., we investigated a new approach to GBM treatment that would overcome the brain's immune response, using mouse NSC and human MSC encapsulated in hyaluronic acid based sECM. In mouse models of human GBM resection, we found that sECM encapsulation of hMSC and mNSC increased their retention in the tumor resection cavity, permitted tumor-selective migration and release of diagnostic and therapeutic proteins *in vivo* (Figure 4). This study demonstrates the efficacy of encapsulated therapeutic stem cells in mouse models of GBM resection and may have implications for developing effective therapies for GBM [157]. In the same study, we investigated the therapeutic potential of sECM encapsulated human bone marrow derived-MSC, expressing S-TRAIL in mouse resection models of GBM. In sECM hMSC-GBM8 co-culture settings, we observed that encapsulated MSC expressing S-TRAIL migrated out of sECMs and tracked

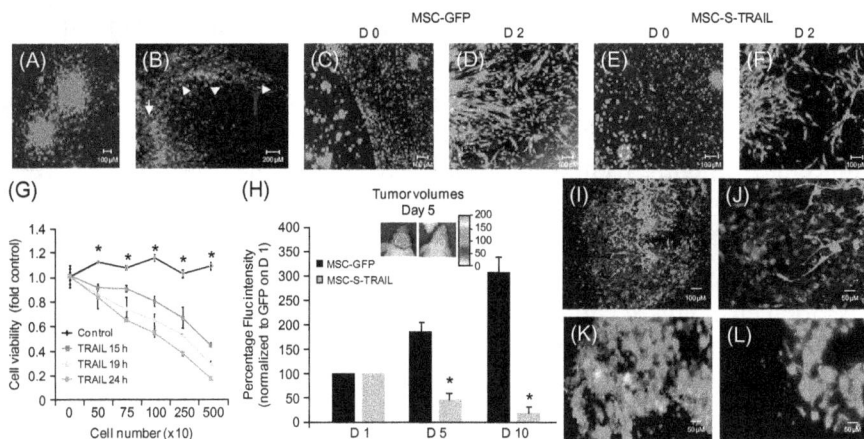

*Figure 4 sECM encapsulated therapeutic human MSC have antitumor effects in primary invasive human GBMs in vitro and in vivo. (A, B) Photomicrographs of primary invasive GBM8-mCherry-Fluc grown as neurospheres in a collagen matrix (A) and serial brain section of mice bearing GBM-mCherry-Fluc tumors showing highly invasive nature of GBM8 (B). Arrows indicate site of implantation and arrowheads indicate path of invasion (B). (C–G) MSC expressing GFP or S-TRAIL were encapsulated in sECMs and placed in the culture dish containing human GBM8-Fluc-mCherry cells. MSC were followed for migration out of sECMs and GBM8 cells were followed for their response to S-TRAIL secreted by MSC. Photomicrographs showing sECM encapsulated hMSC on the day of plating (C, E) and 48 h post-hMSC encapsulation and plating (D, F). (G) Plot showing the GBM8 cell viability at different time points after co-culturing with either sECM encapsulated MSC-GFP or MSC-S-TRAIL (p < 0.05 versus TRAIL). (H–J) Encapsulated MSC-S-TRAIL or MSC-GFP in sECM were implanted intracranially in the tumor resection cavity of mice bearing GBM8-mCherry-Fluc and mice were followed for changes in tumor volume by serial Fluc bioluminescence imaging and correlative immunohistochemistry. Plot and representative figures show the relative mean Fluc signal intensity of sECMs encapsulated MSC-GFP-Fluc or MSC-S-TRAIL bearing mice (tumor volumes: *p < 0.05 versus controls) (H). (I, J) Low (I) and high (J) magnification photomicrographs from the serial brain sections of mice showing MSC (green) on day 5 mice post-MSC implantation in the GBM8 (red) resection cavity. (K, L) Representative images showing cleaved caspase-3 staining (purple) on brain sections from MSC-S-TRAIL (K) and control (L) mice 5 days post-treatment. (For interpretation of the references to color in this figure legend, the reader is referred to the web version of this book.) Adapted from Ref. [157] with permission.*

GBM8 cells and also induced tumor cell apoptosis. Furthermore, to assess the therapeutic potential of sECM encapsulated hMSC-S-TRAIL in mouse resection models of primary GBMs, we tested sECM encapsulated hMSC-S-TRAIL in a mouse GBM8 model of tumor resection cavity. Our results showed a significant decrease in residual GBM8 cells, as well as a sustained presence of encapsulated hMSC in the tumor resection cavity, and MSC migration to invading glioma cells [157].

7. *IN VIVO* IMAGING OF STEM CELL FATE, ANTITUMOR AGENT PHARMACOKINETICS AND THERAPEUTIC EFFICACY

The clinical translation of MSC-based therapies will depend on the development of noninvasive high resolution *in vivo* imaging technologies that can simultaneously monitor the long-term fate of MSC, the pharmacokinetics of MSC-delivered therapeutics, and ultimately the therapeutic efficacy of MSC *in vivo*. An effective approach to visualizing and tracking the fate of stem cells *in vivo* is noninvasive molecular imaging technology [158]. Molecular imaging is a relatively new strategy which aims to visualize and quantify biological, physiological, and pathological processes using various molecular-targeted imaging probes, specific for cell surface biomarkers that are unique to specific cells. This technique has advantages over traditional methods, as it permits real-time tracking *in vivo*.

The successful use of reporter gene imaging is a powerful approach to monitor live cells *in vivo*. A number of developments have emerged over the past few years in high resolution *in vivo* imaging methods, such as radionuclide imaging, that include positron emission tomography (PET) and SPECT, as well as magnetic resonance (MR) imaging and spectroscopy. Another prominent technique is optical imaging, which uses different physical parameters of light interaction with tissue. It relies on fluorescence, absorption, reflectance, or bioluminescence as a source of contrast. Optical imaging techniques that have gained popularity in the last decade include near infrared fluorescence (NIRF) imaging, fluorescence-mediated tomography, bioluminescence imaging, and intravital microscopy [159].

7.1 Optical Imaging

Optical imaging techniques, including bioluminescence and fluorescence imaging, of reporter genes (e.g. green fluorescence protein or luciferase) can be used to access stem cell fate and survival. Bioluminescence (also called chemiluminescence) imaging requires a reporter construct to effect production of a protein called luciferase, an enzyme that provides imaging contrast by the light emission that results from the luciferase-catalyzed conversion of D-luciferin to oxyluciferin. Our laboratory has constructed hybrid reporter constructs of bioluminescence and intravital imaging that allow following the entire process of tumor formation, MSC migration, MSC dispersion throughout the tumor, and MSC killing of glioma cells monitored noninvasively in a longitudinal fashion [60,160−163]. The engineering of MSC with these hybrid reporters also allows real-time imaging of MSC migration, and also the ability to visualize tumor penetration by

MSC at the single cell level. Although, MSC and other stem cell types are promising therapeutic delivery vehicles, preclinical and clinical applications of stem cell-based therapy would benefit from the ability to simultaneously determine therapeutic efficacy and pharmacokinetics of therapies delivered by engineered stem cells.

The creation of new molecules that concurrently enhance tumor cell killing and permit diagnostic tracking are vital to overcoming the limitations rendering current therapeutic regimens for cancers ineffective. We recently investigated the efficacy of an innovative new multifunctional-targeted anticancer molecule, SM7L, using models of GBM. Designed using predictive computer modeling, SM7L incorporates the therapeutic activity of the promising antitumor cytokine MDA-7/IL-24, an enhanced secretory domain, and diagnostic domain for noninvasive tracking [164]. *In vitro* assays revealed the diagnostic domain of SM7L produced robust photon emission, while the therapeutic domain showed marked antitumor efficacy and significant modulation of p38MAPK and ERK pathways. *In vivo*, the unique multifunctional nature of SM7L allowed simultaneous real-time monitoring of both SM7L delivery and antitumor efficacy. Utilizing engineered stem cells as novel delivery vehicles for SM7L therapy (SC-SM7L), we demonstrate that SC-SM7L significantly improved pharmacokinetics and attenuated progression of established peripheral and intracranial human GBM xenografts. Furthermore, SC-SM7L antitumor efficacy was augmented *in vitro* and *in vivo* by concurrent activation of caspase-mediated apoptosis induced by adjuvant SC-mediated S-TRAIL delivery [164]. These studies provide a novel example of the benefits of simultaneous diagnostic tracking and stem cell-delivered therapeutics, using the optimized therapeutic molecule SM7L.

Moreover, in a past study we engineered and screened numerous fusion variants that contained therapeutic (TRAIL) and diagnostic (luciferase) domains designed to allow simultaneous investigation of multiple events in stem cell-based therapy *in vivo*. When various stem cell lines were engineered with the optimized molecule, SRL_OL_2TR, diagnostic imaging showed marked differences in the levels and duration of secretion between stem cell lines, while the therapeutic activity of the molecule showed the different secretion levels translated to significant variability in tumor cell killing [161]. *In vivo*, simultaneous diagnostic and therapeutic monitoring revealed that stem cell-based delivery significantly improved pharmacokinetics and antitumor effectiveness of the therapy compared to intravenous or intratumoral delivery [161]. As a treatment for highly malignant brain tumor xenografts, tracking SRL_OL2TR showed stable stem cell-mediated delivery that significantly regressed peripheral and intracranial tumors. Together, the integrated diagnostic and therapeutic properties of SRL_OL2TR answer critical questions necessary for successful utilization of stem cells as novel therapeutic vehicles.

7.2 Other Imaging Techniques

Given the need for stem cell imaging techniques in larger animals or humans (where bioluminescent imaging is precluded by limited depth of tissue penetration), magnetic resonance imaging (MRI) and PET have been evaluated for feasibility of stem cell tracking. MRI technology is swiftly advancing beyond anatomical imaging to dynamic, metabolic imaging, allowing the investigation of multiple aspects of tumor cell biology and assessment of therapeutic response. In a recent study, MSC labeled with fluorescent magnetic nanoparticles (FMNP) were utilized to access the therapeutic efficacy and targeted imaging of MSC in animal models of gastric tumor. The *in vivo* distribution of FMNP-labeled MSC was observed using fluorescence imaging and MRI system [165]. Other stem cell types have been tracked using this technology, such as in a study by Bakhru et al. in which adult mouse NSC were successfully labeled with ultrasmall superparamagnetic iron oxide nanoparticles that were surface functionalized with water-soluble chitosan and then fluorescently labeled. It was discovered that labeling the NSCs using this strategy enabled nanoparticle-based tracking studies and determination of fate beyond uptake by fluorescence microscopy, as well as tracking of labeled cells as localized regions of hypointensity in MRI images [166].

Clinically, radionuclide imaging, including PET and SPECT, has been routinely used to assess numerous physiological functions, such as cell tracking. PET imaging depicts the biological activity of molecular events and is particularly well suited for the assessment of early changes in disease pathophysiology. In this imaging modality, cells are either labeled directly with a radionuclide or by reporter gene labeling [167]. In the previously mentioned study from our lab in which S-TRAIL and HSV-TK were utilized to treat mouse models of GBM, the systemic administration of GCV post-tumor treatment and subsequent selective elimination of therapeutic MSC expressing HSV-TK *in vitro* and *in vivo* was monitored in real time by PET imaging utilizing 18F-FHBG, a substrate for HSV-TK [101]. In a similar study, Hasenbach and colleagues studied the glioma tropism of bone marrow-derived stem cells by utilizing reporter gene labeling. In this study, the HSV-1-TK gene was introduced into the cells as a reporter gene for PET studies using the tracer 18F-FHBG. Results showed an increased tracer uptake in gliomas following FHBG administration and imaging. These results were further confirmed using a multimodal PET/MRI system and also validated using 2-photon laser scanning microscopy [168]. Radionuclide imaging techniques have been fruitful in tracking stem cells *in vivo* and have also been combined with other imaging modalities to form a multimodal imaging approach that can serves as a great tool in today's imaging world.

8. PROSPECTS AND CAVEATS

Based on the fact that both endogenous and *ex vivo* cultured MSC specifically home to, integrate with, and survive in tumors, a substantial progress has been made in the field of MSC-based therapeutics for cancer. MSC are the most attractive candidates for cell-based therapies in humans as they are easy to isolate and expand *in vitro*, nonimmunogenic, relatively easy to manipulate *in vitro*, have the ability to interact with different tissue environments, and preferentially migrate toward local and disseminated malignant disease. Although bone marrow, adipose, and local tissues are excellent sources of autologous and allogeneic MSC, their regulation and kinetics remain poorly understood. The advantage of autologous MSC is mainly their immunological compatibility, which has a profound effect on cell survival after transplantation. A number of preclinical studies wherein MSC have been used to express a single therapeutic agent for localized delivery have been tested in different cancer types. The use of MSC in combinatorial therapies, such as incorporating MSC-based therapies with traditional drug therapies, delivery of multiple therapeutic agents in an effort to target multiple pathways in cancerous cells with one MSC preparation, or modification of both MSC homing and expression of therapeutic agents, is critical to move MSC-based cancer therapies into clinics. The safety of the grafted MSC is a major concern in a clinical setting.

Importantly, nonimmortalized adult stem cells do not confer the same danger as immortalized adult stem cells and may be used without posing risk to the patient. A number of clinical trials utilizing MSC have not reported any major adverse events from allogeneic transplants [169–171]. A number of ongoing clinical studies using MSC have been listed in Table 1. Autologous bone marrow MSC treatment in transplant patients with subclinical rejection and interstitial fibrosis was reported to be clinically feasible and safe, and was also suggestive of systemic immunosuppression in a recently completed Phase I clinical study [172]. It is important to mention that a few reports have implicated MSC in promoting the growth of certain cancers. However, there have not been significant subsequent reports on the promotion of tumor growth by MSC in recent years. A thorough understanding of MSC biology and fate in tumor models that closely mimic clinical settings are critical when developing MSC-based therapies for clinical translation in cancer patients.

Table 1 Clinical Trials Using Stem Cells for Cancer

Title of Study	Condition	Current Status
Stem Cell Transplant for High Risk Central Nervous System (CNS) Tumors	Glioblastoma	Phase II
Chemotherapy with CD133+ Select Autologous Hematopoietic Stem Cells for Children with Solid Tumors and Lymphomas	Neuroblastoma	Phase I
Mini-Allogeneic Peripheral Blood Progenitor Cell Transplantation for Recurrent or Metastatic Breast Cancer	Breast cancer	Phase II
Chemotherapy Followed by Allogeneic Stem Cell Transplantation for Hematologic Malignancies	Hematologic malignancies	Phase II
High-Dose Chemotherapy with Autologous Stem Cell Rescue in Pediatric High-Risk Brain Tumors	Glioma	Phase II
Combination Chemotherapy Followed by Peripheral Stem Cell Transplant in Treating Young Patients with Newly Diagnosed Supratentorial Primitive Neuroectodermal Tumors or High-Risk Medulloblastoma	Medulloblastoma	Phase III
Mesenchymal Stem Cells in Cisplatin-Induced Acute Renal Failure in Patients with Solid Organ Cancers	Solid tumors	Phase I
Umbilical Cord Blood Transplant for Children with Lymphoid Hematological Malignancies	Lymphoid malignancies	Phase II
Allogeneic Stem Cell Transplant for Chronic Lymphocytic Leukemia	Chronic lymphocytic leukemia	Phase I/II
Autologous Transplant for Multiple Myeloma	Multiple myeloma	Phase II/III

ACKNOWLEDGMENTS

We would like to thank Deepak Bhere and Tracy Twombly for help with the preparation of this manuscript.

REFERENCES

[1] Siegel R, DeSantis C, Virgo K, Stein K, Mariotto A, Smith T, et al. Cancer treatment and survivorship statistics. CA Cancer J Clin 2012;62(4):220−41.

[2] Corsten MF, Shah K. Therapeutic stem-cells for cancer treatment: hopes and hurdles in tactical warfare. Lancet Oncol 2008;9(4):376−84.

[3] Teo AK, Vallier L. Emerging use of stem cells in regenerative medicine. Biochem J 2010;428 (1):11−23.

[4] Yu J, Vodyanik MA, Smuga-Otto K, Antosiewicz-Bourget J, Frane JL, Tian S, et al. Induced pluripotent stem cell lines derived from human somatic cells. Science 2007;318 (5858):1917−20.

[5] Smith S, Neaves W, Teitelbaum S. Adult stem cell treatments for diseases? Science 2006;313 (5786):439.

[6] Friedenstein AJ, Deriglasova UF, Kulagina NN, Panasuk AF, Rudakowa SF, Luria EA, et al. Precursors for fibroblasts in different populations of hematopoietic cells as detected by the in vitro colony assay method. Exp Hematol 1974;2(2):83−92.

[7] Barry FP, Murphy JM. Mesenchymal stem cells: clinical applications and biological characterization. Int J Biochem Cell Biol 2004;36(4):568−84.

[8] Tian H, Bharadwaj S, Liu Y, Ma PX, Atala A, Zhang Y, et al. Differentiation of human bone marrow mesenchymal stem cells into bladder cells: potential for urological tissue engineering. Tissue Eng Part A 2010;16(5):1769−79.

[9] Schipani E, Kronenberg HM. Adult mesenchymal stem cells. StemBook [Internet]. Cambridge (MA) Harvard Stem Cell Institute; 2008-2009, Jan 31.

[10] Sacchetti B, Funari A, Michienzi S, Di Cesare S, Piersanti S, Saggio I, et al. Self-renewing osteoprogenitors in bone marrow sinusoids can organize a hematopoietic microenvironment. Cell 2007;131(2):324−36.

[11] Crisan M, Yap S, Casteilla L, Chen CW, Corselli M, Park TS, et al. A perivascular origin for mesenchymal stem cells in multiple human organs. Cell Stem Cell 2008;3(3):301−13.

[12] Mendez-Ferrer S, Michurina TV, Ferraro F, Mazloom AR, Macarthur BD, Lira SA, et al. Mesenchymal and haematopoietic stem cells form a unique bone marrow niche. Nature 2010;466(7308):829−34.

[13] Gage FH. Mammalian neural stem cells. Science 2000;287(5457):1433−8.

[14] Weissman IL. Translating stem and progenitor cell biology to the clinic: barriers and opportunities. Science 2000;287(5457):1442−6.

[15] Robinton DA, Daley GQ. The promise of induced pluripotent stem cells in research and therapy. Nature 2012;481(7381):295−305.

[16] Weinacht KG, Brauer PM, Felgentreff K, Devine A, Gennery AR, Giliani S, et al. The role of induced pluripotent stem cells in research and therapy of primary immunodeficiencies. Curr Opin Immunol 2012;24(5):617−24.

[17] Shah K. Mesenchymal stem cells engineered for cancer therapy. Adv Drug Deliv Rev 2012;64(8):739−48.

[18] Romanova ON. Clinical features and treatment of polypous rhinosinusitis combined with allergic diseases. Vestn Otorinolaringol 2003;(1):32−4.

[19] Momin EN, Mohyeldin A, Zaidi HA, Vela G, Quinones-Hinojosa A. Mesenchymal stem cells: new approaches for the treatment of neurological diseases. Curr Stem Cell Res Ther 2010;5(4):326−44.

[20] Fukuchi Y, Nakajima H, Sugiyama D, Hirose I, Kitamura T, Tsuji K, et al. Human placenta-derived cells have mesenchymal stem/progenitor cell potential. Stem Cells 2004;22 (5):649−58.

[21] da Silva Meirelles L, Chagastelles PC, Nardi NB. Mesenchymal stem cells reside in virtually all post-natal organs and tissues. J Cell Sci 2006;119(Pt 11):2204−13.

[22] Anderson DJ, Gage FH, Weissman IL. Can stem cells cross lineage boundaries? Nat Med 2001;7(4):393−5.

[23] Jiang Y, Jahagirdar BN, Reinhardt RL, Schwartz RE, Keene CD, Ortiz-Gonzalez XR, et al. Pluripotency of mesenchymal stem cells derived from adult marrow. Nature 2002;418(6893):41−9.

[24] Orlic D, Kajstura J, Chimenti S, Jakoniuk I, Anderson SM, Li B, et al. Bone marrow cells regenerate infarcted myocardium. Nature 2001;410(6829):701−5.

[25] Bentzon JF, Stenderup K, Hansen FD, Schroder HD, Abdallah BM, Jensen TG, et al. Tissue distribution and engraftment of human mesenchymal stem cells immortalized by human telomerase reverse transcriptase gene. Biochem Biophys Res Commun 2005;330(3):633−40.

[26] Kern S, Eichler H, Stoeve J, Kluter H, Bieback K. Comparative analysis of mesenchymal stem cells from bone marrow, umbilical cord blood, or adipose tissue. Stem Cells 2006;24 (5):1294−301.

[27] Mueller SM, Glowacki J. Age-related decline in the osteogenic potential of human bone marrow cells cultured in three-dimensional collagen sponges. J Cell Biochem 2001;82 (4):583−90.

[28] Stewart MC, Stewart AA. Mesenchymal stem cells: characteristics, sources, and mechanisms of action. Vet Clin North Am Equine Pract 2011;27(2):243−61.

[29] Prindull G, Ben-Ishay Z, Ebell W, Bergholz M, Dirk T, Prindull B, et al. CFU-F circulating in cord blood. Blut 1987;54(6):351−9.

[30] Erices A, Conget P, Minguell JJ. Mesenchymal progenitor cells in human umbilical cord blood. Br J Haematol 2000;109(1):235−42.

[31] Goodwin HS, Bicknese AR, Chien SN, Bogucki BD, Quinn CO, Wall DA, et al. Multilineage differentiation activity by cells isolated from umbilical cord blood: expression of bone, fat, and neural markers. Biol Blood Marrow Transplant 2001;7(11):581−8.

[32] Chang YJ, Shih DT, Tseng CP, Hsieh TB, Lee DC, Hwang SM, et al. Disparate mesenchyme-lineage tendencies in mesenchymal stem cells from human bone marrow and umbilical cord blood. Stem Cells 2006;24(3):679−85.

[33] Bieback K, Kern S, Kluter H, Eichler H. Critical parameters for the isolation of mesenchymal stem cells from umbilical cord blood. Stem Cells 2004;22(4):625−34.

[34] Lee OK, Kuo TK, Chen WM, Lee KD, Hsieh SL, Chen TH, et al. Isolation of multipotent mesenchymal stem cells from umbilical cord blood. Blood 2004;103(5):1669−75.

[35] Pierdomenico L, Bonsi L, Calvitti M, Rondelli D, Arpinati M, Chirumbolo G, et al. Multipotent mesenchymal stem cells with immunosuppressive activity can be easily isolated from dental pulp. Transplantation 2005;80(6):836−42.

[36] Shi S, Robey PG, Gronthos S. Comparison of human dental pulp and bone marrow stromal stem cells by cDNA microarray analysis. Bone 2001;29(6):532−9.

[37] Miura M, Gronthos S, Zhao M, Lu B, Fisher LW, Robey PG, et al. SHED: stem cells from human exfoliated deciduous teeth. Proc Natl Acad Sci USA 2003;100(10):5807–12.

[38] He Q, Wan C, Li G. Concise review: multipotent mesenchymal stromal cells in blood. Stem Cells 2007;25(1):69–77.

[39] Hong HS, Lee J, Lee E, Kwon YS, Ahn W, Jiang MH, et al. A new role of substance P as an injury-inducible messenger for mobilization of CD29(+) stromal-like cells. Nat Med 2009;15(4):425–35.

[40] Spaeth EL, Kidd S, Marini FC. Tracking inflammation-induced mobilization of mesenchymal stem cells. Methods Mol Biol 2012;904:173–90.

[41] Spaeth E, Kidd S, Marini FC. Inflammation and tumor microenvironments: defining the migratory itinerary of mesenchymal stem cells. Gene Ther 2008;15(10):730–8.

[42] Momin EN, Vela G, Zaidi HA, Quinones-Hinojosa. The oncogenic potential of mesenchymal stem cells in the treatment of cancer: directions for future research. Curr Immunol Rev 2010;6(2):137–48.

[43] Imitola J, Raddassi K, Park KI, Mueller FJ, Nieto M, Teng YD, et al. Directed migration of neural stem cells to sites of CNS injury by the stromal cell-derived factor 1alpha/CXC chemokine receptor 4 pathway. Proc Natl Acad Sci USA 2004;101(52):18117–22.

[44] Nakamizo A, Marini F, Amano T, Khan A, Studeny M, Gumin J, et al. Human bone marrow-derived mesenchymal stem cells in the treatment of gliomas. Cancer Res 2005;65(8):3307–18.

[45] Son BR, Marquez-Curtis LA, Kucia M, Wysoczynski M, Turner AR, Ratajczak J, et al. Migration of bone marrow and cord blood mesenchymal stem cells in vitro is regulated by stromal-derived factor-1-CXCR4 and hepatocyte growth factor-c-met axes and involves matrix metalloproteinases. Stem Cells 2006;24(5):1254–64.

[46] Quante M, Tu SP, Tomita H, Gonda T, Wang SS, Takashi S, et al. Bone marrow-derived myofibroblasts contribute to the mesenchymal stem cell niche and promote tumor growth. Cancer Cell 2011;19(2):257–72.

[47] Smith H, Whittall C, Weksler B, Middleton J. Chemokines stimulate bidirectional migration of human mesenchymal stem cells across bone marrow endothelial cells. Stem Cells Dev 2012;21(3):476–86.

[48] Darash-Yahana M, Gillespie JW, Hewitt SM, Chen YY, Maeda S, Stein I, et al. The chemokine CXCL16 and its receptor, CXCR6, as markers and promoters of inflammation-associated cancers. PLoS One 2009;4(8):e6695.

[49] Curiel TJ, Coukos G, Zou L, Alvarez X, Cheng P, Mottram P, et al. Specific recruitment of regulatory T cells in ovarian carcinoma fosters immune privilege and predicts reduced survival. Nat Med 2004;10(9):942–9.

[50] Jacobs JF, Idema AJ, Bol KF, Grotenhuis JA. de Vries IJ. Wesseling P, et al. Prognostic significance and mechanism of Treg infiltration in human brain tumors. J Neuroimmunol 2010;225(1-2):195–9.

[51] Jagadeesha DK, Takapoo M, Banfi B, Bhalla RC, Miller Jr. FJ. Nox1 transactivation of epidermal growth factor receptor promotes N-cadherin shedding and smooth muscle cell migration. Cardiovasc Res 2012;93(3):406–13.

[52] Maretzky T, Evers A, Zhou W, Swendeman SL, Wong PM, Rafii S, et al. Migration of growth factor-stimulated epithelial and endothelial cells depends on EGFR transactivation by ADAM17. Nat Commun 2011;2:229.

[53] Yahata Y, Shirakata Y, Tokumaru S, Yang L, Dai X, Tohyama M, et al. A novel function of angiotensin II in skin wound healing. Induction of fibroblast and keratinocyte migration by angiotensin II via heparin-binding epidermal growth factor (EGF)-like growth factor-mediated EGF receptor transactivation. J Biol Chem 2006;281(19):13209–16.

[54] Porcile C, Bajetto A, Barbieri F, Barbero S, Bonavia R, Biglieri M, et al. Stromal cell-derived factor-1alpha (SDF-1alpha/CXCL12) stimulates ovarian cancer cell growth through the EGF receptor transactivation. Exp Cell Res 2005;308(2):241–53.

[55] Roelle S, Grosse R, Aigner A, Krell HW, Czubayko F, Gudermann T. Matrix metalloproteinases 2 and 9 mediate epidermal growth factor receptor transactivation by gonadotropin-releasing hormone. J Biol Chem 2003;278(47):47307–18.

[56] De Becker A, Van Hummelen P, Bakkus M, Vande Broek I, De Wever J, De Waele M, et al. Migration of culture-expanded human mesenchymal stem cells through bone marrow endothelium is regulated by matrix metalloproteinase-2 and tissue inhibitor of metalloproteinase-3. Haematologica 2007;92(4):440–9.

[57] Qiu Y, Marquez-Curtis LA, Janowska-Wieczorek A. Mesenchymal stromal cells derived from umbilical cord blood migrate in response to complement C1q. Cytotherapy 2012;14 (3):285–95.

[58] Francois S, Bensidhoum M, Mouiseddine M, Mazurier C, Allenet B, Semont A, et al. Local irradiation not only induces homing of human mesenchymal stem cells at exposed sites but promotes their widespread engraftment to multiple organs: a study of their quantitative distribution after irradiation damage. Stem Cells 2006;24(4):1020–9.

[59] Aboody KS, Brown A, Rainov NG, Bower KA, Liu S, Yang W, et al. Neural stem cells display extensive tropism for pathology in adult brain: evidence from intracranial gliomas. Proc Natl Acad Sci USA 2000;97(23):12846–51.

[60] Sasportas LS, Kasmieh R, Wakimoto H, Hingtgen S. van de Water JA, Mohapatra G, et al. Assessment of therapeutic efficacy and fate of engineered human mesenchymal stem cells for cancer therapy. Proc Natl Acad Sci USA 2009;106(12):4822–7.

[61] Wei J, Blum S, Unger M, Jarmy G, Lamparter M, Geishauser A, et al. Embryonic endothelial progenitor cells armed with a suicide gene target hypoxic lung metastases after intravenous delivery. Cancer Cell 2004;5(5):477–88.

[62] Maestroni GJ, Hertens E, Galli P. Factor(s) from nonmacrophage bone marrow stromal cells inhibit Lewis lung carcinoma and B16 melanoma growth in mice. Cell Mol Life Sci 1999;55(4):663–7.

[63] Nakamura K, Ito Y, Kawano Y, Kurozumi K, Kobune M, Tsuda H, et al. Antitumor effect of genetically engineered mesenchymal stem cells in a rat glioma model. Gene Ther 2004;11 (14):1155–64.

[64] Qiao C, Xu W, Zhu W, Hu J, Qian H, Yin Q, et al. Human mesenchymal stem cells isolated from the umbilical cord. Cell Biol Int 2008;32(1):8–15.

[65] Qiao L, Xu Z, Zhao T, Zhao Z, Shi M, Zhao RC, et al. Suppression of tumorigenesis by human mesenchymal stem cells in a hepatoma model. Cell Res 2008;18(4):500–7.

[66] Khakoo AY, Pati S, Anderson SA, Reid W, Elshal MF. Rovira, II, et al. Human mesenchymal stem cells exert potent antitumorigenic effects in a model of Kaposi's sarcoma. J Exp Med 2006;203(5):1235–47.

[67] Otsu K, Das S, Houser SD, Quadri SK, Bhattacharya S, Bhattacharya J. Concentration-dependent inhibition of angiogenesis by mesenchymal stem cells. Blood 2009;113 (18):4197–205.

[68] Gondi CS, Veeravalli KK, Gorantla B, Dinh DH, Fassett D, Klopfenstein JD, et al. Human umbilical cord blood stem cells show PDGF-D-dependent glioma cell tropism in vitro and in vivo. Neuro Oncol 2010;12(5):453–65.

[69] Dasari VR, Kaur K, Velpula KK, Dinh DH, Tsung AJ, Mohanam S, et al. Downregulation of focal adhesion kinase (FAK) by cord blood stem cells inhibits angiogenesis in glioblastoma. Aging (Albany NY) 2010;2(11):791–803.

[70] Dasari VR, Kaur K, Velpula KK, Gujrati M, Fassett D, Klopfenstein JD, et al. Upregulation of PTEN in glioma cells by cord blood mesenchymal stem cells inhibits migration via downregulation of the PI3K/Akt pathway. PLoS One 2010;5(4):e10350.

[71] Dasari VR, Velpula KK, Kaur K, Fassett D, Klopfenstein JD, Dinh DH, et al. Cord blood stem cell-mediated induction of apoptosis in glioma downregulates X-linked inhibitor of apoptosis protein (XIAP). PLoS One 2010;5(7):e11813.

[72] Jiao H, Guan F, Yang B, Li J, Shan H, Song L, et al. Human umbilical cord blood-derived mesenchymal stem cells inhibit C6 glioma via downregulation of cyclin D1. Neurol India 2011;59(2):241−7.

[73] Akimoto K, Kimura K, Nagano M, Takano S. To'a Salazar G, Yamashita T, et al. Umbilical cord blood-derived mesenchymal stem cells inhibit, but adipose tissue-derived mesenchymal stem cells promote, glioblastoma multiforme proliferation. Stem Cells Dev 2013;22(9):1370−86.

[74] Okada H, Pollack IF. Cytokine gene therapy for malignant glioma. Expert Opin Biol Ther 2004;4(10):1609−20.

[75] Zhang X, Zhang L, Xu W, Qian H, Ye S, Zhu W, et al. Experimental therapy for lung cancer: umbilical cord-derived mesenchymal stem cell-mediated interleukin-24 delivery. Curr Cancer Drug Targets 2013;13(1):92−102.

[76] Chen X, Lin X, Zhao J, Shi W, Zhang H, Wang Y, et al. A tumor-selective biotherapy with prolonged impact on established metastases based on cytokine gene-engineered MSCs. Mol Ther 2008;16(4):749−56.

[77] Gao P, Ding Q, Wu Z, Jiang H, Fang Z. Therapeutic potential of human mesenchymal stem cells producing IL-12 in a mouse xenograft model of renal cell carcinoma. Cancer Lett 2010;290(2):157−66.

[78] Seo SH, Kim KS, Park SH, Suh YS, Kim SJ, Jeun SS, et al. The effects of mesenchymal stem cells injected via different routes on modified IL-12-mediated antitumor activity. Gene Ther, 2011;18(5):488−95.

[79] Ryu CH, Park SH, Park SA, Kim SM, Lim JY, Jeong CH, et al. Gene therapy of intracranial glioma using interleukin 12-secreting human umbilical cord blood-derived mesenchymal stem cells. Hum Gene Ther 2011;22(6):733−43.

[80] Xu X, Yang G, Zhang H, Prestwich GD. Evaluating dual activity LPA receptor pan-antagonist/autotaxin inhibitors as anti-cancer agents in vivo using engineered human tumors. Prostaglandins Other Lipid Mediat 2009;89(3-4):140−6.

[81] Hu W, Wang J, Dou J, He X, Zhao F, Jiang C, et al. Augmenting therapy of ovarian cancer efficacy by secreting IL-21 human umbilical cord blood stem cells in nude mice. Cell Transplant 2011;20(5):669−80.

[82] Chawla-Sarkar M, Leaman DW, Borden EC. Preferential induction of apoptosis by interferon (IFN)-beta compared with IFN-alpha2: correlation with TRAIL/Apo2L induction in melanoma cell lines. Clin Cancer Res 2001;7(6):1821−31.

[83] Johns TG, Mackay IR, Callister KA, Hertzog PJ, Devenish RJ, Linnane AW. Antiproliferative potencies of interferons on melanoma cell lines and xenografts: higher efficacy of interferon beta. J Natl Cancer Inst 1992;84(15):1185−90.

[84] Wong VL, Rieman DJ, Aronson L, Dalton BJ, Greig R, Anzano MA. Growth-inhibitory activity of interferon-beta against human colorectal carcinoma cell lines. Int J Cancer 1989;43(3):526−30.

[85] Studeny M, Marini FC, Champlin RE, Zompetta C, Fidler IJ, Andreeff M. Bone marrow-derived mesenchymal stem cells as vehicles for interferon-beta delivery into tumors. Cancer Res 2002;62(13):3603−8.

[86] Studeny M, Marini FC, Dembinski JL, Zompetta C, Cabreira-Hansen M, Bekele BN, et al. Mesenchymal stem cells: potential precursors for tumor stroma and targeted-delivery vehicles for anticancer agents. J Natl Cancer Inst 2004;96(21):1593–603.

[87] Ren C, Kumar S, Chanda D, Chen J, Mountz JD, Ponnazhagan S. Therapeutic potential of mesenchymal stem cells producing interferon-alpha in a mouse melanoma lung metastasis model. Stem Cells 2008;26(9):2332–8.

[88] Ren C, Kumar S, Chanda D, Kallman L, Chen J, Mountz JD, et al. Cancer gene therapy using mesenchymal stem cells expressing interferon-beta in a mouse prostate cancer lung metastasis model. Gene Ther 2008;15(21):1446–53.

[89] Dembinski JL, Wilson SM, Spaeth EL, Studeny M, Zompetta C, Samudio I, et al. Tumor stroma engraftment of gene-modified mesenchymal stem cells as anti-tumor therapy against ovarian cancer. Cytotherapy 2013;15(1):20–32: e2.

[90] Wang GX, Zhan YA, Hu HL, Wang Y, Fu B. Mesenchymal stem cells modified to express interferon-beta inhibit the growth of prostate cancer in a mouse model. J Int Med Res 2012;40(1):317–27.

[91] Bitsika V, Roubelakis MG, Zagoura D, Trohatou O, Makridakis M, Pappa KI, et al. Human amniotic fluid-derived mesenchymal stem cells as therapeutic vehicles: a novel approach for the treatment of bladder cancer. Stem Cells Dev 2012;21(7):1097–111.

[92] Grander D, Einhorn S. Interferon and malignant disease—how does it work and why doesn't it always? Acta Oncol 1998;37(4):331–8.

[93] Lens M. Cutaneous melanoma: interferon alpha adjuvant therapy for patients at high risk for recurrent disease. Dermatol Ther 2006;19(1):9–18.

[94] Sartoris S, Mazzocco M, Tinelli M, Martini M, Mosna F, Lisi V, et al. Efficacy assessment of interferon-alpha-engineered mesenchymal stromal cells in a mouse plasmacytoma model. Stem Cells Dev 2011;20(4):709–19.

[95] Danks MK, Yoon KJ, Bush RA, Remack JS, Wierdl M, Tsurkan L, et al. Tumor-targeted enzyme/prodrug therapy mediates long-term disease-free survival of mice bearing disseminated neuroblastoma. Cancer Res 2007;67(1):22–5.

[96] Altaner C. Prodrug cancer gene therapy. Cancer Lett 2008;270(2):191–201.

[97] Miletic H, Fischer Y, Litwak S, Giroglou T, Waerzeggers Y, Winkeler A, et al. Bystander killing of malignant glioma by bone marrow-derived tumor-infiltrating progenitor cells expressing a suicide gene. Mol Ther 2007;15(7):1373–81.

[98] Choi SA, Lee JY, Wang KC, Phi JH, Song SH, Song J, et al. Human adipose tissue-derived mesenchymal stem cells: characteristics and therapeutic potential as cellular vehicles for prodrug gene therapy against brainstem gliomas. Eur J Cancer 2012;48(1):129–37.

[99] Song C, Xiang J, Tang J, Hirst DG, Zhou J, Chan KM, et al. Thymidine kinase gene modified bone marrow mesenchymal stem cells as vehicles for antitumor therapy. Hum Gene Ther 2011;22(4):439–49.

[100] Cavarretta IT, Altanerova V, Matuskova M, Kucerova L, Culig Z, Altaner C. Adipose tissue-derived mesenchymal stem cells expressing prodrug-converting enzyme inhibit human prostate tumor growth. Mol Ther 2010;18(1):223–31.

[101] Martinez-Quintanilla J., Bhere D., Heidari P., He D., Mahmood U., Shah K. In vivo imaging of the therapeutic efficacy and fate of bimodal engineered stem cells in malignant brain tumors. Stem Cells, 2013.

[102] Aghi M, Martuza RL. Oncolytic viral therapies—the clinical experience. Oncogene 2005;24 (52):7802–16.

[103] Parato KA, Senger D, Forsyth PA, Bell JC. Recent progress in the battle between oncolytic viruses and tumours. Nat Rev Cancer 2005;5(12):965–76.

[104] Nakashima H, Kaur B, Chiocca EA. Directing systemic oncolytic viral delivery to tumors via carrier cells. Cytokine Growth Factor Rev 2010;21(2-3):119−26.

[105] Power AT, Bell JC. Cell-based delivery of oncolytic viruses: a new strategic alliance for a biological strike against cancer. Mol Ther 2007;15(4):660−5.

[106] Pereboeva L, Komarova S, Mikheeva G, Krasnykh V, Curiel DT. Approaches to utilize mesenchymal progenitor cells as cellular vehicles. Stem Cells 2003;21(4):389−404.

[107] Komarova S, Kawakami Y, Stoff-Khalili MA, Curiel DT, Pereboeva L. Mesenchymal progenitor cells as cellular vehicles for delivery of oncolytic adenoviruses. Mol Cancer Ther 2006;5(3):755−66.

[108] Stoff-Khalili M.A., Rivera A.A., Mathis J.M., Banerjee N.S., Moon A.S., Hess A., et al. Mesenchymal stem cells as a vehicle for targeted delivery of CRAds to lung metastases of breast carcinoma. Breast Cancer Res Treat, 2007.

[109] Sonabend AM, Ulasov IV, Tyler MA, Rivera AA, Mathis JM, Lesniak MS. Mesenchymal stem cells effectively deliver an oncolytic adenovirus to intracranial glioma. Stem Cells 2008;26(3):831−41.

[110] Yong RL, Shinojima N, Fueyo J, Gumin J, Vecil GG, Marini FC, et al. Human bone marrow-derived mesenchymal stem cells for intravascular delivery of oncolytic adenovirus Delta24-RGD to human gliomas. Cancer Res 2009;69(23):8932−40.

[111] Garcia-Castro J, Alemany R, Cascallo M, Martinez-Quintanilla J, Arriero Mdel M, Lassaletta A, et al. Treatment of metastatic neuroblastoma with systemic oncolytic virotherapy delivered by autologous mesenchymal stem cells: an exploratory study. Cancer Gene Ther 2010;17(7):476−83.

[112] Mader EK, Butler G, Dowdy SC, Mariani A, Knutson KL, Federspiel MJ, et al. Optimizing patient derived mesenchymal stem cells as virus carriers for a phase I clinical trial in ovarian cancer. J Transl Med 2013;11:20.

[113] Ong H.T., Federspiel M.J., Guo C.M., Lucien Ooi L., Russell S.J., Peng K.W., et al. Systemically delivered measles virus-infected mesenchymal stem cells can evade host immunity to inhibit liver cancer growth. J Hepatol, 2013.

[114] Jain RK, di Tomaso E, Duda DG, Loeffler JS, Sorensen AG, Batchelor TT. Angiogenesis in brain tumours. Nat Rev Neurosci 2007;8(8):610−22.

[115] Ma X, Zhang L, Ma H, Yu R. Association of vascular endothelial growth factor expression with angiogenesis and tumor cell proliferation in human lung cancer. Zhonghua Nei Ke Za Zhi 2001;40(1):32−5.

[116] Ghaedi M, Soleimani M, Taghvaie NM, Sheikhfatollahi M, Azadmanesh K, Lotfi AS, et al. Mesenchymal stem cells as vehicles for targeted delivery of anti-angiogenic protein to solid tumors. J Gene Med 2011;13(3):171−80.

[117] Folkman J, Watson K, Ingber D, Hanahan D. Induction of angiogenesis during the transition from hyperplasia to neoplasia. Nature 1989;339(6219):58−61.

[118] Samant RS, Shevde LA. Recent advances in anti-angiogenic therapy of cancer. Oncotarget 2011;2(3):122−34.

[119] Zheng L, Zhang D, Chen X, Yang L, Wei Y, Zhao X. Antitumor activities of human placenta-derived mesenchymal stem cells expressing endostatin on ovarian cancer. PLoS One 2012;7(7):e39119.

[120] Batchelor TT, Sorensen AG, di Tomaso E, Zhang WT, Duda DG, Cohen KS, et al. AZD2171, a pan-VEGF receptor tyrosine kinase inhibitor, normalizes tumor vasculature and alleviates edema in glioblastoma patients. Cancer Cell 2007;11(1):83−95.

[121] Kadambi A, Mouta Carreira C, Yun CO, Padera TP, Dolmans DE, Carmeliet P, et al. Vascular endothelial growth factor (VEGF)-C differentially affects tumor vascular function

and leukocyte recruitment: role of VEGF-receptor 2 and host VEGF-A. Cancer Res 2001;61(6):2404−8.

[122] Tong RT, Boucher Y, Kozin SV, Winkler F, Hicklin DJ, Jain RK. Vascular normalization by vascular endothelial growth factor receptor 2 blockade induces a pressure gradient across the vasculature and improves drug penetration in tumors. Cancer Res 2004;64(11):3731−6.

[123] Hormigo A, Gutin PH, Rafii S. Tracking normalization of brain tumor vasculature by magnetic imaging and proangiogenic biomarkers. Cancer Cell 2007;11(1):6−8.

[124] Bexell D, Gunnarsson S, Tormin A, Darabi A, Gisselsson D, Roybon L, et al. Bone marrow multipotent mesenchymal stroma cells act as pericyte-like migratory vehicles in experimental gliomas. Mol Ther 2009;17(1):183−90.

[125] Walczak H, Krammer PH. The CD95 (APO-1/Fas) and the TRAIL (APO-2L) apoptosis systems. Exp Cell Res 2000;256(1):58−66.

[126] Mueller L.P., Luetzkendorf J., Widder M., Nerger K., Caysa H., Mueller T. TRAIL-transduced multipotent mesenchymal stromal cells (TRAIL-MSC) overcome TRAIL resistance in selected CRC cell lines in vitro and in vivo. Cancer Gene Ther, 2010.

[127] Kim SK, Cargioli TG, Machluf M, Yang W, Sun Y, Al-Hashem R, et al. PEX-producing human neural stem cells inhibit tumor growth in a mouse glioma model. Clin Cancer Res 2005;11(16):5965−70.

[128] Ehtesham M, Kabos P, Gutierrez MA, Chung NH, Griffith TS, Black KL, et al. Induction of glioblastoma apoptosis using neural stem cell-mediated delivery of tumor necrosis factor-related apoptosis-inducing ligand. Cancer Res 2002;62(24):7170−4.

[129] Ehtesham M, Kabos P, Kabosova A, Neuman T, Black KL, Yu JS. The use of interleukin 12-secreting neural stem cells for the treatment of intracranial glioma. Cancer Res 2002;62 (20):5657−63.

[130] Loebinger MR, Eddaoudi A, Davies D, Janes SM. Mesenchymal stem cell delivery of TRAIL can eliminate metastatic cancer. Cancer Res 2009;69(10):4134−42.

[131] Grisendi G, Bussolari R, Cafarelli L, Petak I, Rasini V, Veronesi E, et al. Adipose-derived mesenchymal stem cells as stable source of tumor necrosis factor-related apoptosis-inducing ligand delivery for cancer therapy. Cancer Res 2010;70(9):3718−29.

[132] Ciavarella S, Grisendi G, Dominici M, Tucci M, Brunetti O, Dammacco F, et al. In vitro anti-myeloma activity of TRAIL-expressing adipose-derived mesenchymal stem cells. Br J Haematol 2012;157(5):586−98.

[133] Shah K, Bureau E, Kim DE, Yang K, Tang Y, Weissleder R, et al. Glioma therapy and real-time imaging of neural precursor cell migration and tumor regression. Ann Neurol 2005;57(1):34−41.

[134] Shah K, Tung CH, Yang K, Weissleder R, Breakefield XO. Inducible release of TRAIL fusion proteins from a proapoptotic form for tumor therapy. Cancer Res 2004;64 (9):3236−42.

[135] Ozdemir V, Williams-Jones B, Glatt SJ, Tsuang MT, Lohr JB, Reist C. Shifting emphasis from pharmacogenomics to theragnostics. Nat Biotechnol 2006;24(8):942−6.

[136] Noble M. Can neural stem cells be used as therapeutic vehicles in the treatment of brain tumors? Nat Med 2000;6(4):369−70.

[137] Nesterenko I, Wanningen S, Bagci-Onder T, Anderegg M, Shah K. Evaluating the effect of therapeutic stem cells on TRAIL resistant and sensitive medulloblastomas. PLoS One 2012;7(11):e49219.

[138] Kim SM, Woo JS, Jeong CH, Ryu CH, Lim JY, Jeun SS. Effective combination therapy for malignant glioma with TRAIL-secreting mesenchymal stem cells and lipoxygenase inhibitor MK886. Cancer Res 2012;72(18):4807−17.

[139] Corsten MF, Miranda R, Kasmieh R, Krichevsky AM, Weissleder R, Shah K. MicroRNA-21 knockdown disrupts glioma growth in vivo and displays synergistic cytotoxicity with neural precursor cell delivered S-TRAIL in human gliomas. Cancer Res 2007;67(19):8994−9000.

[140] Bagci-Onder T., Wakimoto H., Anderegg M., Cameron C., Shah K. A dual PI3K/mTOR inhibitor, PI-103, cooperates with stem cell delivered TRAIL in experimental glioma models. Cancer Res, 2010.

[141] Mohr A, Albarenque SM, Deedigan L, Yu R, Reidy M, Fulda S, et al. Targeting of XIAP combined with systemic mesenchymal stem cell-mediated delivery of sTRAIL ligand inhibits metastatic growth of pancreatic carcinoma cells. Stem Cells 2010;28(11):2109−20.

[142] Kim SM, Oh JH, Park SA, Ryu CH, Lim JY, Kim DS, et al. Irradiation enhances the tumor tropism and therapeutic potential of tumor necrosis factor-related apoptosis-inducing ligand-secreting human umbilical cord blood-derived mesenchymal stem cells in glioma therapy. Stem Cells 2010;28(12):2217−28.

[143] Yu R, Deedigan L, Albarenque SM, Mohr A, Zwacka RM. Delivery of sTRAIL variants by MSCs in combination with cytotoxic drug treatment leads to p53-independent enhanced antitumor effects. Cell Death Dis 2013;4:e503.

[144] Yulyana Y, Endaya BB, Ng WH, Guo CM, Hui KM, Lam PY, et al. Carbenoxolone enhances TRAIL-induced apoptosis through the upregulation of death receptor 5 and inhibition of gap junction intercellular communication in human glioma. Stem Cells Dev 2013;22(13):1870−82.

[145] Van de Water JA, Bagci-Onder T, Agarwal AS, Wakimoto H, Kasmieh R, Roovers RC, et al. Therapeutic stem cells expressing different variants of EGFR-specific nanobodies have anti-tumor effects. 109(41):16642−7.

[146] Kim SW, Kim SJ, Park SH, Yang HG, Kang MC, Choi YW, et al. Complete regression of metastatic renal cell carcinoma by multiple injections of engineered mesenchymal stem cells expressing dodecameric TRAIL and HSV-TK. Clin Cancer Res 2013;19(2):415−27.

[147] Yan C, Li S, Li Z, Peng H, Yuan X, Jiang L, et al. Human umbilical cord mesenchymal stem cells as vehicles of CD20-specific TRAIL fusion protein delivery: a double-target therapy against non-Hodgkin's lymphoma. Mol Pharm 2013;10(1):142−51.

[148] Murua A, Portero A, Orive G, Hernandez RM, de Castro M, Pedraz JL. Cell microencapsulation technology: towards clinical application. J Control Release 2008;132(2):76−83.

[149] Morris PJ. Immunoprotection of therapeutic cell transplants by encapsulation. Trends Biotechnol 1996;14(5):163−7.

[150] Rihova B. Immunocompatibility and biocompatibility of cell delivery systems. Adv Drug Deliv Rev 2000;42(1-2):65−80.

[151] Pan L, Ren Y, Cui F, Xu Q. Viability and differentiation of neural precursors on hyaluronic acid hydrogel scaffold. J Neurosci Res 2009;87(14):3207−20.

[152] Park KI, Teng YD, Snyder EY. The injured brain interacts reciprocally with neural stem cells supported by scaffolds to reconstitute lost tissue. Nat Biotechnol 2002;20(11):1111−7.

[153] Teng YD, Lavik EB, Qu X, Park KI, Ourednik J, Zurakowski D, et al. Functional recovery following traumatic spinal cord injury mediated by a unique polymer scaffold seeded with neural stem cells. Proc Natl Acad Sci USA 2002;99(5):3024−9.

[154] Sayyar B, Dodd M, Wen J, Ma S, Marquez-Curtis L, Janowska-Wieczorek A, et al. Encapsulation of factor IX-engineered mesenchymal stem cells in fibrinogen-alginate microcapsules enhances their viability and transgene secretion. J Tissue Eng 2012;3(1): 2041731412462018.

[155] Reagan MR, Seib FP, McMillin DW, Sage EK, Mitsiades CS, Janes SM, et al. Stem cell implants for cancer therapy: TRAIL-expressing mesenchymal stem cells target cancer cells in situ. J Breast Cancer 2012;15(3):273−82.

[156] Muldoon LL, Soussain C, Jahnke K, Johanson C, Siegal T, Smith QR, et al. Chemotherapy delivery issues in central nervous system malignancy: a reality check. J Clin Oncol 2007;25(16):2295–305.

[157] Kauer TM, Figueiredo JL, Hingtgen S, Shah K. Encapsulated therapeutic stem cells implanted in the tumor resection cavity induce cell death in gliomas. Nat Neurosci 2012;15 (2):197–204.

[158] Xia T, Jiang H, Li C, Tian M, Zhang H. Molecular imaging in tracking tumor stem-like cells. J Biomed Biotechnol 2012;420364.

[159] Kurland BF, Gerstner ER, Mountz JM, Schwartz LH, Ryan CW, Graham MM, et al. Promise and pitfalls of quantitative imaging in oncology clinical trials. Magn Reson Imaging 2012;30(9):1301–12.

[160] Bagci-Onder T, Wakimoto H, Anderegg M, Cameron C, Shah K. A dual PI3K/mTOR inhibitor, PI-103, cooperates with stem cell-delivered TRAIL in experimental glioma models. Cancer Res 2011;71(1):154–63.

[161] Hingtgen SD, Kasmieh R. van de Water J, Weissleder R, Shah K. A novel molecule integrating therapeutic and diagnostic activities reveals multiple aspects of stem cell-based therapy. Stem Cells 2010;28(4):832–41.

[162] Shah K. Imaging neural stem cell fate in mouse model of glioma. Curr Protoc Stem Cell Biol 2009; [Chapter 5: p. Unit 5A 1].

[163] Shah K, Hingtgen S, Kasmieh R, Figueiredo JL, Garcia-Garcia E, Martinez-Serrano A, et al. Bimodal viral vectors and in vivo imaging reveal the fate of human neural stem cells in experimental glioma model. J Neurosci 2008;28(17):4406–13.

[164] Hingtgen S, Kasmieh R, Elbayly E, Nesterenko I, Figueiredo JL, Dash R, et al. A first-generation multi-functional cytokine for simultaneous optical tracking and tumor therapy. PLoS One 2012;7(7):e40234.

[165] Ruan J, Ji J, Song H, Qian Q, Wang K, Wang C, et al. Fluorescent magnetic nanoparticle-labeled mesenchymal stem cells for targeted imaging and hyperthermia therapy of in vivo gastric cancer. Nanoscale Res Lett 2012;7(1):309.

[166] Bakhru SH, Altiok E, Highley C, Delubac D, Suhan J, Hitchens TK, et al. Enhanced cellular uptake and long-term retention of chitosan-modified iron-oxide nanoparticles for MRI-based cell tracking. Int J Nanomedicine 2012;7:4613–23.

[167] Chan AT, Abraham MR. SPECT and PET to optimize cardiac stem cell therapy. J Nucl Cardiol 2012;19(1):118–25.

[168] Hasenbach K, Wiehr S, Herrmann C, Mannheim J, Cay F, von Kurthy G, et al. Monitoring the glioma tropism of bone marrow-derived progenitor cells by 2-photon laser scanning microscopy and positron emission tomography. Neuro Oncol 2012;14(4):471–81.

[169] Jang YO, Kim YJ, Baik SK, Kim MY, Eom YW, Cho MY, et al. Histological improvement following administration of autologous bone marrow-derived mesenchymal stem cells for alcoholic cirrhosis: a pilot study. Liver Int 2013.

[170] Jiang PC, Xiong WP, Wang G, Ma C, Yao WQ, Kendell SF, et al. A clinical trial report of autologous bone marrow-derived mesenchymal stem cell transplantation in patients with spinal cord injury. Exp Ther Med 2013;6(1):140–6.

[171] Wang S, Cheng H, Dai G, Wang X, Hua R, Liu X, et al. Umbilicalcord mesenchymal stem cell transplantation significantly improves neurological function in patients with sequelae of traumatic brain injury. Brain Res 2013;1532:76–84.

[172] Reinders ME, de Fijter JW, Roelofs H, Bajema IM, de Vries DK, Schaapherder AF, et al. Autologous bone marrow-derived mesenchymal stromal cells for the treatment of allograft rejection after renal transplantation: results of a phase I study. Stem Cells Transl Med 2013;2(2):107–11.

www.ingramcontent.com/pod-product-compliance
Lightning Source LLC
Chambersburg PA
CBHW060514220326
41598CB00025B/3657